# Beyond The Theories of Newton, Maxwell and others

## Ramesh L. Joshi, Ph.D., P.E.

ISBN: 978-1-365-14410-3

This Book is dedicated to Jekorba G.Trivedi, my grandmother, and all grandmothers who nurture and love their grandchildren.
You are noble.

# Contents

# Preface

For a change in *space*, *time*, and *matter*
of a *physical event* has *a position*,
but must have *connection*.

## Why?

Because continuity and discontinuity
transcend in our finite understanding;

the latter demands individuality and
freedom;

the former establishes connection between
the individuals and unions;

and the process of dealing with them
together provides progress in science.

This book is about what we have observed daily and yet not incorporated in the great intellectual adventures and moral quests of Isaac Newton (1642-1742), Michael Faraday (1791-1867), James Clerk Maxwell (1831-1879), and others in the search for the laws of motion of materials under the gravitation and electromagnetic fields. The book answers the fundamental questions: How one can go beyond the established classical theories when scientists are working on the advanced quantum mechanics and quantum electrodynamics? What treasures can one find in the natural physical sciences to improve these well-established theories?

I had the good fortune of having many caring, enthusiastic teachers at all levels of my education. In particular, I had excellent professors during my study at graduate levels. In particular: Professor Dahayabhai Motibhai Patel (circa 1900-1983), Sardar Patel University, Vallabh Vidyanagar, India; Professors Prahladbhai Chunilal Vaidya (1918-2010) and Darshan Singh Bassan (1927-1999), both at the Gujarat University, Ahmedabad, India, who taught mathematical foundations of dynamics, electricity, and magnetism, and theory of relativity. They clearly showed the foundations, logically applicable freedom on the rules in erecting the natural structure of the subjects, with less concern about its practical applications, inspired to look for the nature around us, and observe and learn the order and harmony of the universe. On the other hand, Professors Hans Albert Einstein (1904-1973) and Paul Lieber (1918-1992), both at the University of California, Berkeley taught the physical science with the applications to the practical foundations of classical mechanics, fluid mechanics, and geometry. They were concerned with the vast range of problems arising in science and current technology, but had less emphasis on the available freedom in mathematics and physics; they showed how the laws of nature work in governing the phenomenon under study and their applications to daily life.

All of them taught of the need to spend time to understand the fundamentals and raise questions such as: Why something works in a specific way and not the other way around? I sat near them as an Indian student and saw through the mathematical reasoning that Newtonian theory does not explain the two sides of earth—the phenomena of day and night occurring in its revolution around the sun. I sat with them as an American student and saw through a parity of professors and pupils, the union of matter and space, the union of space and time (and without an arbitrary introduction of the hypothesis of constancy of velocity of light), spinning of electrons, and earth. All these facts occur in our everyday observations, but had no explanation in current physics or mathematics. Thus, they instructed me to see beyond the giants' known theories.

These late professors also taught lessons about life. First, these theories are well-established and it will be hard to improve easily and get any financial support as it is available for the new developments in physics and mathematics. So be prepared to work without expecting instant results of immediate rewards, and so I continue to tinker for a long time without getting frustrated. Second, when I fully understand the fundamentals and shake the foundations, new results will follow.

During the last few years after my retirement, I got guidance and moral support from Professor George Leitmann, 91 (1925- ), also at the University of California, Berkeley to work on my findings. He suggested me to work with and follow the reasoning of the old theories, but to look for the new approaches.

And through the chats during my school days with Professor Hans Albert Einstein, ideas indirectly I received from his mother—the first wife of Albert Einstein—Mileva Marić Einstein (1875-1948), and practical inspiration from my grandmother, Jekorba G. Trivedi (circa 1890-1982). They always compelled not to play with prod probing into the well-established fundamentals of social and physical sciences to produce holes in anyone of them. They encouraged doing what is right even if it is risky, rather than what is popular, or easily doable, or essentially acceptable at the time. Grandmothers Jekorba and Mileva taught me: An observer (with a reference frame) observing an individual, indivisible entity, whether a material entity or a human being has identity of mass (M), has position in space (S), and has association with and related to time (T)—has activities in space and time, and associated with the matters; whether the entity is a small one (a particle or a peasant) or a large one (a planet or a president), has its identity and its position, and one should respect them both. We need to be aware of these two characteristics—physical identity and geometrical position—and consider them separate where required; these characteristics are influenced by and influencing to connect with each other and during its motion (work) may unite with space, time, and matter (social, political or otherwise related to its work).

This book tells the story of various physical objects (entities) with their geometrical positions in space, their motions in time and their associated matter, and their mutual connections and their unions where it takes place. This is the strength of the book; it reveals the limitations of the classical theories, and shows how to remove them successfully taking us beyond the classical mechanics of Newton, and electrodynamics of Faraday, Maxwell and others, without entering into the special theory of relativity, general relativity, quantum mechanics, and quantum electrodynamics developed in the twentieth century.

The book is intended to explain a few of the naturally occurring phenomena—not explained by the classical mechanics and electrodynamics. The work presumes the reader is familiar with basic knowledge of Newtonian laws of mechanics and some familiarity with electricity and magnetism, and basic mathematics. Despite the shortness of the book, the reader will be benefited by simply reading, while ignoring the equations if required, to understand the basic existing mysteries not explained in the studies of physics, chemistry, and mathematics presented at the undergraduate college level courses.

The book is written in the shaky San Francisco Bay Area, the State of California, to shake the foundations of the classical mechanics and electrodynamics. Though the Bay Area shakes, it has built the wonderful Silicon Valley with many high tech industries and institutions, and has advanced the technology in the last fifty years far faster and better than what was done over five hundred years, and we all are benefited throughout the world. Similarly, from the fundamentals of the theories presented in the book, we will shake it, but will not shift it; will stand on it, but will not stance on it; will transcend to see beyond what we know and open the avenues of uncovering the deeper secrets of nature that will easily explain: why the earth (planet) spins but not its moon. Realize that if the earth is not spinning, we will not have day and night; if the moon is spinning, we would have seen the backside (far-side) of it, and NASA did not need to go to verify the back (far-side) side of moon in the late 1960s. Similarly, there are explanations on the observed facts, like: why velocity of light is constant; why electrons spin; how to represent ½ spin of electrons; why are there seven rows and eight columns in the periodic table of atomic elements.

I am thankful to my professors for their suggestions and guidance to develop the results presented in this book. All the credits for the results go to my professors and grandmothers; if there is any blame, it is mine.

Ramesh L. Joshi

# Introduction

In the physical world,
*changes are fundamental*,
and *time* permits a *change.*

## Why?

**Because scientists like to talk about changes associated with physical entities and events so much that a change has become a cliché of everyday use: "*Change is constant,*" and it is fundamental in the study of science.**

A change is not a passive activity; it is fully active starting from its occurrence until it finishes. An entity participates in the action of change with its relocation in space (S), and in its alteration in mass or matter (M). There are plenty of examples of change we observe every day, every second in our lives. A laminar water flow in a river, or the rapid and turbulent river flow—the kind of deadly even for professional rafters who approach gingerly—are examples in which water changes its locations and materials in flow of water.

In physics and mathematics, the change is usually expressed as motion of material, or flow of material, displaced from one location to the next in space, and change in quantity of material. For a given motion—either a spatial one or a material one under a limiting case—one can always find the change to be uniform (or may be non-uniform). The uniformity is expressed as time. One of the examples of motions of driving in the congested traffic of San Francisco Bay-Area is the distance defined in terms of time and not in terms of actual length from a point (say, home) to the final destination (say, office), and people say: "I live 15 minutes away from office." This is the way we have united the distance and time on one side and defined time on the other.

Scientists have used this uniform change to introduce the unit of time as is done for the spatial and material units. This is the way units of space, time, and matter came in science and everyday life of mankind. And the order and analysis of the changes lead to physical theories.

Physical theories are developed to understand the changes, the motions, occurring in the nature or the performance of the planned experiments and explain how these phenomena work. These theories permit scientists to make predictions to lead to the new insights into the naturally occurring phenomena. The theories which are simple—defined in terms of clear assumptions—easily understood, and can be applied to different realms of experiences and unite them are considered good theories. Good physical theory survives under the constantly changing physical world. The experimental facts associated with a theory are undisputable. When experimental data are not accommodated by theory, the theory is held in question. For the good theories, scientists do not easily abandon the theory, like when Charles Augustin Coulomb (1736-1806) introduced the concept of attraction and repulsion between two charges to follow the inverse square law of Newton between two bodies; or work hard to explain what is going on with the theory like how Faraday worked with Newtonian concept in developing to understand properties of charged particles but introduced the concept of field. In the final investigation for the physical theory, the experimental results, once tested and retested, verified by the independent experimental methods, the experimental result ultimately overrules the theory.

As noted above, one of the troublemakers in physical theories is charged particles—electrons. Electrons and their associated fields are everywhere in science, in art, in computer, in communication, and in all aspects of physical, social and scientific worlds. They are in the space—occupied by the matter or not occupied by the matter known as empty space; in time—the past, the present or the future; and in matter—the solid, liquid and also in the vapor form of mass or that appears in the form of energy. The space, time, and matter—all of them contain influence from electrons or their fields; the movements of matter at a given time in an observed space have direct cause and effect from electrons. One of the goals of physical science is to understand the motions and activities associated with the electrons that explain how the universe works.

Since the time of Johannes Kepler (1571-1630), Galileo Galilei (1564-1642), and Newton, we have focused on observable large bodies, considered them as particles, influencing each other to have forces at a distance, represented in our knowledge of natural phenomena to explain planetary motions and other physical laws. Newton extended the laws of dynamics to understand motion of very small particles of light—corpuscular theory of light—and explained optical phenomena and properties of light known at that time. Because of Newton's overwhelming scientific prestige, the corpuscular theory of light was well accepted and remained in use until the beginning of the nineteenth century, though wave theory of light introduced by Christian Huygens (1629-1695) at the end of seventeenth century (1690) before Newton first published his work of "Optics" (in 1704).

The precedent of noted particle theory continued in development of Coulomb's law of electrical charges of the same and opposite signs. Faraday, Maxwell, and Heinrich Rudolf Hertz (1857-1894) extended the motion of charged particles, and brought that their motion representing the electromagnetic waves.

They had no problem to develop the above noted particle theories that establish a connection between the very large and very small, but when it came to creating harmony with optical wave theory, controversy has persisted since the time of Newton, and continued with Max Karl Ernst Ludwig Planck (1858-1947), Einstein and others, and still it is not over. Under these circumstances, the charged particles—electrons—have made large contributions but have been considered as subordinate participant in science, second class citizen as if neither having very large, nor very small size so that they did not get the same status of particles or the waves. Due to these limitations, we will go over and see how the electrons tacitly provide the needed forces to have major contribution in science, and will reveal the unknown phenomena and overlooked facts of classical mechanics and electrodynamics. Fortunately, looking at the classical theory developed for the charged particles, we will report that it has all the required characteristics of particles and waves to compromise the controversy and more to reveal those not reported in classical science.

A lion cub will compromise by remaining hungry, but will not eat grass to violate its natural instinct of having nutrition from higher food chain. In the same way, electrons have compromised with the particles and waves theories, but have not resolved the differences or settled with originating an alternate inferior theory. Where it is not feasible to compromise, electrons have raised red flags of creating a new theory like Special theory of relativity or strong doubts like uncertainty principle in quantum mechanics, but have not violated with the higher standards of physics. We will see here that electrons and their fields have always provided larger properties in science including the one that will combine both of them, reveal the scientific facts not reported in Newton's mechanics and Maxwell's electrodynamics.

Both the theories have started with the essential properties of the media they represent to provide laws of mechanics and optics to explain the particle motions and wave interactions with the media. The driving forces for both the theories lie in the interaction of particles and waves in the given space. Newton's gravitational field depends how the two bodies are interacting at a distance in a given space. In the wave theory how the wave surface is moving in the given media and its effect is carried out. The foundation of both the theories is based on the connection of matter to act with other matter in a given space and observed time.

The word electron is a relatively newcomer in science. Joseph John Thomson (1856-1940) discovered it in 1897 to explain the experimental observations of the motion of positively- and negatively-charged particles in understanding the structure of atom. We will focus on charged particles—electrons—and their fields' contribution in the past four centuries in science.

Electrons are locally connected with space as a point (of Newtonian particle), surface and volume, without requiring a commutative medium to influence other electrons as their associated fields accomplish the requirements. By connection we mean: A geometrical representation of matter associated (with the Newtonian particles and electrons) in space at a given time during its motion, and its interaction with other matter.

The fundamental difference between the matter of Newtonian particle and matter of charged particle (electron) is that, during the study of its motion, the former can be reduced to a point, while the latter cannot be, except in the study of Coulomb's law which considers them as points to follow with Newtonian particle theory. Newtonian bodies and planets appear as particles, points in the space; space, time, and matter remain absolute, and forces are acting at a distance. Charged particles are treated as particles, but their associated fields appear in the form of surfaces and volumes, and act in their neighborhood. Due to these fundamental properties, these scientific factual properties have dictated in the representation of motion of the gravitational particles in dynamics, and charged particles in electrodynamics, respectively, through ordinary differential and partial differential equations.

In this book we will discuss the electron's motions studied under classical electrodynamics. We will neither consider the results from the special theory of relativity nor from quantum mechanics.

As noted in the Preface, for changes associated with entities, we will establish connection with space, time, and matter of Newtonian particles associated with large bodies and reveal why planets spin but not their moons, which is excluded from the Newton's gravitational laws; the waves appearing in Maxwell's equations and establish the union of space and matter, and observe the union of space and time already included in it, that appeared before the birth of special theory of relativity; and study the periodic table of all elements based on the various possible connections of the electrons and explain why it is necessary to have the table to arrange in eight columns and seven rows. In Appendix A we will present an experiment to demonstrate that the magnetic field represents a couple and not a force. The noted experiment also demonstrates that the unpaired electrons in iron, cobalt, and nickel spin and produce the spinning of the u-clips (easy to see, but can be seen with a suspended similar material wire also) in front of magnetic field having the magnetic couple. Also in the Appendix, we will explain and demonstrate how the unpaired electrons in iron, cobalt, and nickel spin.

# Chapter I

## Under gravity
## The Earth and Moon revolve (move) in elliptic orbits, but only earth spins and not its moon.

### Why?

Because gravity on the celestial bodies has different geometrical structures of connections

## Introduction

We are grateful to Newton, Galileo, Kepler, and others for their contribution in understanding how planets, stars, and rigid bodies move. Without their contributions to science we would not have made the progress enjoyed in our lives. But, if you want to have a shining example of the limitations in their theories of motion, you do not need to go any further than examining simple phenomena that occur every day in our lives. We daily observe that the sun rises in the east and sets in the west due to revolutions of the earth on its axis; on the other hand, we cannot see the backside of the moon, only its front face (near side). As a matter of fact, before astronauts went to the moon, scientists needed to find out what was on the backside (far-side) of the moon. So the space agency NASA sent satellites to orbit the moon and collected the data before astronauts were sent. Thereafter, astronauts landed on the observed front side (near-side) of the moon. These facts are neither derivable nor explainable from the Newtonian gravitational theory and its advancement on Einstein's general theory of relativity, and thus have fundamental limitations.

Under the mutual gravitational forces, gravitational theories explain very well how the earth revolves around the sun, and the moon revolves around the earth. The same is the case for other planets and their moons. Under the same gravitational forces, the earth spins as shown in Figure 1.1, but not her moon. The same is the case for other planets and their moons. These gravitational theories cannot explain the spin of the earth and planets. What are the limitations in these theories? Why?

# Elliptic Orbits of Earth and Moon

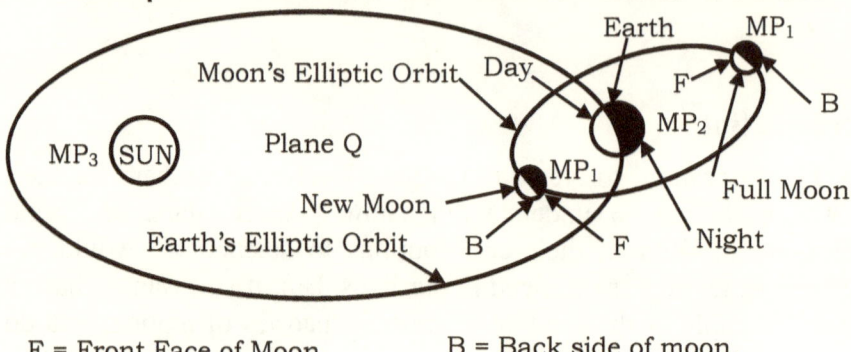

F = Front Face of Moon,
Visible from Earth

B = Back side of moon,
Not visible from Earth

Moon face changes as it revolves around the earth.
Shown only 2 phases of the Moon

$MP_1$, $MP_2$ and $MP_3$ are (3) point-locations of Moon, Earth, and
Sun forming a Plane Q in which earth and moon orbit

Fig. 1.1

These theories of motion, as developed by Galileo and Newton, are based on the concept of particles (considered as points) that include large bodies like the sun, earth, moon, planets, pendulums, and other similar rigid bodies larger than a simple point. The particle (point) has no width, no thickness and no height. The development of the particle (point) theory enjoys the movements of large bodies without any clutter and clatter of points in space in which they are located. The avoidance of clutter appears by restricting the motion having interaction of only two particles without addressing their internal structures, and that of clatter appears by keeping the two particles at a distance. These restrictions are carried on to study the motion of two particles at a distance, and three or more particles need to be reduced into two particles. A study of motion of three particles (bodies) is limited. Under these restrictions, matter of the particles, their locations in space, and their movements with time maintain the space, time, and matter absolute, and they have no connections with each other.

By limiting the motion of the earth and the moon to the restricted concept of particles we cannot simultaneously associate or explain the spin phenomena, though these particles are moving under gravitational influence of three bodies—sun, earth, and moon—and one of them, earth, spins whereas the moon does not.

Many scientists, before me, have been disappointed by the concept of a particle associated with the material in motion. We need to understand the existing definition of a particle and introduce a generalized definition of a particle that permits us to explain the spin phenomena of earth and electrons.

## 1.1 Definition of a particle

When we talk about material in motion, we should be forgiven if we are in the dilemma of not being sure what to call it. Should we call it a particle, or a body, or a wave, or wave-particle? It is an open question for the last four and half centuries, and scientists have yet to come to a concluding agreement. Each scientist has his own choice to label a material with the name he feels comfortable with and gets involved in a discussion with those who do not agree with him. For example, Newton, Einstein, and others liked to study the phenomena of light as particles, while Robert Hooke (1735-1703), Christian Huygens (1629-1695), and others liked to study it as waves. Either way, the phenomena of light remains the same, whether studied as particles or as waves. So, to avoid these confusing conditions, let us first clarify and settle the issue by going to the basics and review why we have different names for the same material in motion.

Kinematics deals with the geometry of motion of points and reference frames, rather than particles, rigid bodies or waves and forces acting on the material. By geometry of motion, we mean the description of position and changes in position of geometrical points with respect to the reference frames. The description is geometrical and the geometry is continuous.

A motion of material, whether it is considered a particle, a body, or a wave, is a description of changes of its position relative to a reference frame. A reference frame can be fixed on earth, on the sun, on a star (for example the North Star), or can be fixed in space. For a selected reference frame, there are a number of different geometrical coordinate systems available to measure the changes in position of the material. A reference frame is purely a kinematical device used for geometrical description of motion without regard to mass or size of the material or forces involved to create motion.

In the description of relative motion of material in space described with respect to a reference frames, the material and the reference frame, both are local and discontinuous, while the geometry of the space, in which the motion takes place, is global and continuous. These are the fundamental differences and a prominent difficulty to accommodate motion of the material under the reaction from other materials and forces without characterizing the material as particle or body or wave.

For the relative motion with respect to the reference frames, Newton overcame the discontinuity by keeping the space and time absolute, and introducing an inertial (reference) frame by defining as the one in which the laws of motion hold good. In order to be precise in our presentation, and to distinguish it from other reference frames moving relative to the inertial frame which we will introduce later after the definition of particle, we will denote the inertial reference frame by K and call it a "stationary system or stationary reference frame." Thus, the stationary reference frame unites with the geometry of the space in which the material is moving and satisfies the laws of motion. And the inertial reference frame establishes geometrical continuity, at least between two geometrical points; one of the points is located at the origin of the reference frame, the second one to be on the material and by considering it to be a particle.

Kinetics, on the other hand, deals with the effect of forces—as Newton considered the action and reaction of earth and the sun by considering the motion of their center of mass as particles, or as Maxwell considered action and reaction of electric charges and magnets by considering the motion of their changes and their associated fields' changes in their neighboring space as bodies, and for the first time Hertz considered these changes to demonstrate the electromagnetic phenomena as waves. To accommodate the particle and wave concepts for the electrical charges and photons, Louis-Victor-Pierre-Raymond De Broglie (1892-1987), inspired by the work of Max Karl Ernst Ludwig Planck (1858-1947) and Einstein, referred to them as wave-particles. De Broglie's wave-particle concept permitted Erwin Schrödinger (1887-1961) to introduce wave equations for quantum mechanics.

Kinetics is the study of the actions and relations of forces that result in the motion of materials. Thus, kinetics deals with the study of changes in geometry, materials, and forces. In this study, we model the material in motion that can act as particles, or as rigid bodies, or as waves, depending on how the matter and the forces are working. And there is no reason to impose *a priori* (word) limitation on the material as a particle, or a body, or a wave in the study of kinetics.

The current dealings of kinetics combines the material points with geometrical points in its motion, without considering that these two items—material and geometry—need to be considered separate entities, and thus it is confusing and imposes restrictions on whether to consider the material as particle, rigid body, or waves, depending upon how the material behaves during the motion under the effect of other materials or forces. For example, in the orbital motion of the earth around the sun, the model of Newtonian mechanics considered both of them as points—particles—since the acceleration (*a*) of the mass center of the earth, based on the collection of all particles of the earth, is proportional to the force caused by the gravitational reaction of the sun and earth on the earth. In this kinetics picture, without recognizing that the two items should be separate, we combined the mass center of the earth material with a geometrical point on its orbit.

The reverse is the case for the charged particles and magnets, in which Maxwell considered them as bodies in derivation of the Maxwell's equations. In this derivation it was necessary that electrical charges be considered as a body allowing him to see how the electric and magnetic fields interact and change in the neighboring space with the motion of charged particles (now named electrons).

Whether to call a clump of material a wave, or a body, or a particle is not that important, as long as the name permits the study of its motion in physics and mathematics. Naming any material as particle, as was the case with Newton, facilitates the study of its changes in space with time. But let us not be confused with using the word particle in a restricted sense that it is limited by having no width, no length, and no height. In reality, the word particle introduces a bridge filling the gap between the material and the space connecting the two sets of points—the material and the geometrical—permitting to study the motion.

Note that the model of material particle introduced here has front and back, top and bottom, right and left sides with respect to K, and thus must have a minimum of two points (and the reasons of two points will follow after the definition of particle) in space at a given time and they both are not necessarily the same. The same way, a geometrical point also has the similar properties, and it is united with the time as required by the constancy of light, (which we will see in Chapter IV that is) required by the Maxwell's equations. So, depending upon what level of study is carried out, a material particle has (a minimum of) two geometrical points, and it associates with other material or geometry or time and we will see that the material particle under study creates additional points to participate in its motion.

To remove the noted limitations, and to permit the current usage of the word particle, used for planets, moon, electrons, photons, and elementary particles, and so on, we define a particle as follows:

*A particle has material points that associate with geometrical points; these two sets of points may be conjugated (combined) or may be separated (discrete) depending on the material characteristics how it manifest with the space and time in its motion.*

Thus, the various forms of matter, the material particles manifest themselves through the geometrical points; in other words, these material points are directly related and naturally dependent on geometrical points and that they are exchangeable with the geometrical points in space and time. At the first observation, particularly in Newtonian mechanics, it is hard to distinguish these two sets of points.

For example, in Newtonian mechanics, one can study the motion of a planet by considering its mass center, with a suitable stationary reference frame K. In this case, the material point and its orbital-geometrical point are combined and are the same. In this case, focusing on only a single point is sufficient to study the motion of the earth, planets, and other large bodies; it is not necessary to have more than a point or a particle for the study of their motions, and no geometrical discontinuity in the motion with respect to the reference frame K.

According to Newton, however, the laws of motion of a particle, let's say P, are also valid under other reference frames, we will represent it as K', as long as it moves with uniform velocity with respect to K. This is a kinematical—geometrical—property applied to the kinetics of motion of particles. To accommodate this case with K', it requires having two geometrical points associated with the particle: one where the mass center of the material is located and the other is in the direction of motion denoting the uniform velocity of the particle with respect to the reference frame K', as the time is absolute for both the reference frames in Newtonian mechanics. The single point P in K and the appearance of two points in K' cause confusion and a discontinuity for the particle P. To remove the confusion and discontinuity associated with P in the reference frame K', as per new and above noted definition of particle, let us consider that particle P has material point and also an associated geometrical point, such that these two points are in agreement with K and K'.

Likewise, for the Newtonian motion, one can have a mass center at a point without having any material at mass center. For examples, one can study the motion of mass center of uniform density concentric hollow sphere, or can study the motion of a mass center of two spheres connected by a mutual attraction or by a rigid weightless rod, or a mass center of doughnut, or annulus or similar other objects. In this case, the mass center is a geometrical point, while the material is located away from the mass center. In these cases, the material and its course of motion point are different, and thus require having two sets of points to consider—material points and a mass center geometrical point which satisfy the Newton's laws of motion.

The defined model of particle is based on the older concept of particle in physics, represented by Newton and Einstein, and has additional geometrical point(s) associated with material that accommodates the principle of relativity for K and K', maintains its geometrical continuity by bringing material and geometrical points on equal status.

Similarly, for the electrically charged particles in Coulomb's law have both the points—material and geometrical points—are combined. For the Maxwell equations these two points are different and, to avoid the confusion, Maxwell called them bodies, as those points change with the space and time, act in the neighboring space, and it is necessary to have the minimum of four points (of a body) to study their motion. A concept of a single point or a single particle is not sufficient.

The above definition of particle permits expanding point concept associated with material and geometrical points into body, waves, and other possibilities, that was not considered or permissible in earlier classical physics; the new definition permits two sets of points to associate with other material points, or with other geometrical points or with time or with their combinations, or with other forces to cause motion of the material. Under the allowed permission of association, one can build the bridge, not only to convert the particle into a line with two points, surface and waves with three points (having curvature and torsion), volume and body with four points; the bridges having multiple points with permissible paths—either observable or can be developed—to have motions for the defined particles. This definition of the particle has neither constrain nor confusion of moving under K or K'.

Based on the new definition of particle, it will be easy to see that the velocity of light is constant in Maxwell electrodynamics, reveal the geometrical properties having curvature and torsion permitted by electrons in motion that satisfy the Maxwell equations. Also only using Maxwell equations, we will explain why we have seven rows and eight periodic groups (columns) of the atomic elements in the periodic table without referring to the spectrum of light appearing in quantum mechanics.

In this book we will consider the particles as having material point(s) and they associate with geometrical point(s) and, in general, these two sets of points are considered separate.

## 1.2   Introduction of connection

The development of spin phenomena relinquishes the limited concept of a particle, a body that occupies a single point in space and requires establishing a connection of a material particle having influence from other particles and geometrical points in motion.

The non-explained spin disorder in Newtonian mechanics is similar to a two party—two points—dispute with limitations not acceptable to one or both parties requiring resolution to permit both the parties to move forward. What do we do in day to day work to breakout from an impasse like this? We do not abandon the relation of two parties, or their entire work, or theory, but consult, salvage the good results and data, and build a structure that removes the restrictions, and proposes a solution acceptable to both the parties, permitting new future parties to join the theory, as Galileo and Newton did in developing their theories by extending the Aristotle's theory of motion. (We will not go back beyond Kepler in this book.) In the presented studied motion of one, two or more particles, the connection is the fundamental property of the study.

At this juncture of the physical theories of motion of (material) particles suggest that Space, Time and Matter (STM) are connected; one needs to look for the generalized concept of connection, which we use in our daily life without addressing it in terms of STM.

The word 'connection' is a dynamic term, used in common conversations, permitting it to be established in terms of space, time, and matter, providing a generalization of the two points of connection used in classical mechanics. There exists a connection between two friends, between two colleagues working together on a project, between two material bodies—points or particles, between two

11

charges, between two points in space, between two particles in motion combining their space and matter connections, between two related events occurring at two different times combining the matter and time, and so on. In this consideration of connection, sometimes it loses identity or imposes restrictions on friends or colleagues working on a common project, or in motion losing identity of point or matter and so on. In this situation, the loss is not recognized, causing the restriction on the motion, limiting the explanation of the possible motions of the matter. We will keep the track of all possible points of connection in motion.

Newtonian mechanics considers body as a particle—all material points in the body are equivalent and consequently one can focus on a single point in the body—and in its motion can be studied as a geometrical point of the particle. In this consideration, all particles appear as one compact conglomerate entity with no separation between two material particles, no restriction and also no distinction in studying the motion of whole body as particle, and no connection with its surrounding space; like one for all and all for one. In this presentation, to put the particles and bodies on the same standing, we will use the word body for a particle and vice versa interchangeably.

In reality, a material particle which occupies a spatial point and a geometrical point in space are different. For the study of Newtonian mechanics, let us be clear, that at a given time, it happens that these two different points associated with space and material coincide, but they two are separate entities. An introduction of a connection with space and matter associated with the particle will make this distinction.

A motion of celestial (gravitational) body depends on its geometrical (spatial) position and its mass, which contributes as single point but does not explicitly participate in the motion. Due to these spatial dependencies of the celestial bodies in Newtonian mechanics, we will introduce their connections first and call it a spatial connection.

The spatial connection of a particle (body) consists of a number of geometrical points having a functional relation among the points. The spatial connection points are: First point of connection is the location point of the material, second set of points denotes the effect from the particle motion in the space that appear as geometrical points, and third set of material particles having geometrical points in the space that contribute to the motion of the particle. In this definition of spatial connection, there is difference between the geometrical points and material points that appear as geometrical points. These two types of geometrical points are different. By keeping this difference in mind, we define a spatial connection of particle (matter) in the space as follows:

> *Spatial Connection of matter is a functional relationship of matter with a number of geometrical and material points in the space that participate in describing its motion.*

These points of connection can be expressed as functional relationship, like a distance between two points associated with the particles. In this chapter we will focus on motion of the gravitational bodies for which, as noted before, space, time and matter are independent and absolute.

The above definition of spatial connection is adequate to study the motion in Newtonian mechanics, but is limited in studying motions of electrons, photons, elementary particles and also to introduce the concept of induction, velocity of light and so on.

In the following chapters we will see that electrons are not limited to a single point of geometrical connection, like the gravitational particles. An electron at rest may have more than one geometrical point of connection at its location, and in its motion it has more than four points of connection which connects with other points of electrons and also that of the space.

The connection for electrons and that of space are not limited to a single (geometrical) point. In general, a connection of an electron has two, three, or more points of connection with space or with other particles, and also has connection with other geometrical points in the space. These points are requisite to introduce connection. For a complete study of motion of electrons and photons, we generalize the notion of connection and define it as follows:

> *Connection is a (geometrical) requisite on Space–Time–Matter (STM) Manifold to have a number of points in space with time to represent motion of matter.*

A connection may be spatial, or temporal, or material, or their combinations, and we will refer to them accordingly. We will use the above noted generalized definition of connection and where possible introduce it as an analytical function—called a spatial connection, or use a geometrical diagram to explain it, or define a particular one for a specific application, like ST (space-time) connection for the velocity of light or SST (spatial-space-time) connection. The ST and SST connections will be introduced in Chapter III, to explain, respectively, the velocity of light, and the induction current introduced by André-Marie Ampere (1775-1836) and Faraday, and so on. To represent a connection in STM, a number of geometrical points are not necessary, but it helps to visualize it.

In the definition of connection, we stated the number of geometrical points as a requisite. To visualize all possible motions, first we address the required number of points as requisite and then discuss their consequences on possible motions.

The space, time, and matter, and their associated manifolds change during motions of particles and bodies. The connection of the motion under study reveals the participating points, associated surfaces and volumes in the changing STM manifolds, and keeps the associated, participating entities, dots connected.

Relinquishing restrictions of Newtonian mechanics in studying the motions of celestial bodies will permit and recognize the spatial connection for the observed spin, and will provide a way to deduce through calculations that the earth spins due to gravitational effects from the moon that itself does not spin under the same field.

## 1.3 Connections of Constant Velocity Particles and Inertial Frames

First, let us consider an example of two friends (bodies) A and B communicating without any interaction between them about the activities they observe in their neighborhood (in their neighboring space), considered to be satisfying the laws of Euclidean geometry. In this case, A and B have their own identities and different locations one from the other. To simplify our discussion, let us consider A is sitting at a point O, and B is at point $P_1$ away from A and walking slowly but steadily from the point $P_1$ in a straight line at uniform velocity. Friend A has its own identity and its location—they are both fixed and are at point O—and this way A has two points of connection, but they coincide. In the same way, B has its own identity $P_1$, but its location is constantly changing with respect to his friend A, or with respect to the location point O. Thus B also has two points of connection—one at his identity point—and the second one B is expressed in terms of the straight line length from point O. In this case, two points of connection for B are located at different points. Thus, friends A and B each have two points of connections. Their surrounding neighborhood observations reported in their communications will be the same, as long as B is moving with the same small, linear, uniform velocity that follows from daily experience, which is known as Galilean Principle of Relativity (GPR) in physics.

Now, A and B have two possibilities in their communications, interactions, and relative motions. Case 1: A and B maintain a constant relative velocity between the two of them, including a special case in which both are sitting at rest. Case 2: B can run and change his velocity to introduce acceleration in his motion. To represent these two cases in mathematical format let us do the following:

Let $K_1$ be a preferred reference frame to describe motion of a particle $P_1$ (Friend B) with time t considered as Newtonian time, which is absolute. Let the Center of Mass (CM) of a body occupy a geometric point $P_1$, and the same is denoted as Material Point $MP_1$ in reference frame $K_1$. Let us consider the friend A is at the origin O of the reference frame.

The $MP_1$ is a material point and the $P_1$ is a geometrical point. Both are different but are at the same location. Let CS be a suitable coordinate system describing the position of $MP_1$ by a vector *X*, with components $X_i$, where (i = 1, 2, 3), in the space. In this book we are considering the space to be Euclidean, and vectors will be denoted in bold Italic letters.

The description of motion of $MP_1$ with the reference frame $K_1$ is described as changes in its position, velocity, acceleration, and its surroundings. Let $V$ be the velocity and $A$ be the acceleration of $MP_1$ as shown in Figure1.2. The position of $MP_1$ and its constant velocity are respectively represented with respect to reference frame $K_1$ as set of two geometrical points (O, $P_1$), and three geometrical points (O, $P_1$, P'). In this description of uniform velocity of $MP_1$, without loss of generality, one can select the origin O of the coordinate system at $P_1$, and the uniform velocity is expressed in terms of two points ($P_1$, P'), one of which is a material point while the other is a geometrical one.

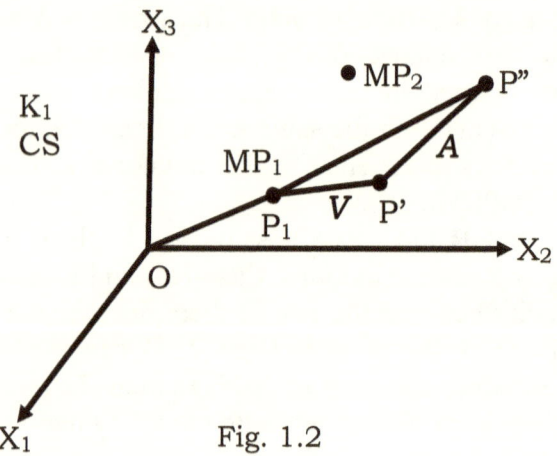

Fig. 1.2

A relative motion of inertial reference frames is a case in which one of the frames is in rectilinear motion with constant uniform velocity and has no effect from any other particle. Galileo's Principle states: A body is in rectilinear uniform motion as long as influence (force) of other bodies does not act on it. As a special case, when uniform velocity is zero, two points $P_1$ and P' coincide.

The spatial points of connection of a particle may be of the same with geometrical points, may be of different geometrical and material points, but are not necessarily always the same. A particle has a geometrical location, which is defined with respect to the origin of the reference frame. Under the uniform motion or under the inertial reference frames, the particle has two points of spatial connection. In these two points of spatial connection, one of them is the material point and the second point is in space; when the body is at rest, they both coincide and one of the spatial points of connection becomes latent. The two-points of spatial connections have no influence from any other particle.

This is a case of two particles, two bodies, or two friends, each having two points of connection in space. They can be at rest, or one of them can be at the origin of the coordinate system, while the other may be moving with uniform relative velocity. Both the entities, at the best represent a straight-line, two points connection in the space, but there is no internal or external force acting or interacting between them.

Thus, based on the Galilean principle, inertial frames of reference and a body or a particle at rest or in rectilinear uniform motion have maximum of two points of connection.

## 1.4 Acceleration under an influence of a force, or another body

In the second case friend (body) B starts running and changes velocity or his direction of motion to introduce acceleration in his activities. This is possible only if there is either an internal force, or is using an external force source to change the velocity or direction of motion. In either case, B (Body) changes its uniform velocity under the force. To associate the case of external force, let us consider B and A are two large bodies of mass $m_1$ and $m_2$ respectively located at $MP_1$ and $MP_2$ originating a force of gravitational attraction $F$ between two of them.

When a particle $MP_1$ is moving under an influence of a force $F$ or other particle, let's say $MP_2$, producing an acceleration $A$ in its motion as shown in Fig. 1.2, the acceleration has three geometrical points of connection ($P_1$, P', P''). These three geometrical points are associated with material $m_1$ and $MP_2$ is influencing its motion. These three points describe acceleration of $MP_1$ and also for $m_1$. We need to be clear that a set of three geometrical points are required to describe acceleration. The acceleration is no more a motion of a point, or that of a particle, but it is a cluster of three geometrical points.

By Newton's second law of motion, the motion of $m_1$ is described by

$$m_1 \boldsymbol{A} = \boldsymbol{F} \text{ or } m_1 A_i = F_i \qquad (1.4.1)$$

In the Equation (1.4.1), the left hand side $m_1$ has a minimum of two points of connection, and its acceleration has three points of connection. In these connections, $m_1$ and its acceleration have a minimum of one common point creating a four-point spatial connection. In the following, we will see that all these four points are not independent, and must have one of the points (in the neighborhood) on a line joining the material $m_1$ and the other material/force causing the acceleration to $m_1$. This dependency makes one of the points of connection latent, and to the spatial connection turns to a three-point spatial connection. Based on this comment, the right hand side force $\boldsymbol{F}$ has three-points of connection.

Newton's universal law of gravitation says that two particles with masses $m_1$ and $m_2$ attract each other with a force. The force of attraction on $m_1$ due to $m_2$ is expressed as $\boldsymbol{F} = \boldsymbol{F}_{12}$, and is given by:

$$\boldsymbol{F} = \boldsymbol{F}_{12} = K \frac{m_1 m_2}{r^2} \underline{e}_r \qquad (1.4.2)$$

In the expression of force $\boldsymbol{F}$, K is a constant; r is the distance between two particles and $\underline{e}_r$ is the unit vector along the line connecting two particles $m_1$ and $m_2$. The $\boldsymbol{F}$ is a function of two independent particles and each particle has two-points of connection giving rise to four-points of connection for the force $\boldsymbol{F}$. We will see in the following that the force $\boldsymbol{F}$ has a constraint to reduce the four-point spatial connection to three-point spatial connection.

Per Newton's third law of motion, the action of $m_2$ on $m_1$ is equal and opposite to reaction of that experienced by $m_1$ and these two forces act along the line joining the two particles. The action and reaction of forces $\boldsymbol{F}$ and $-\boldsymbol{F}$ have four points of connection. Two points are at the material points where masses $m_1$ and $m_2$ are located. The remaining two points are not necessarily in the neighborhood. The action and reaction act along the line joining the two material points. These forces to act along a line impose a constraint on the third (material) point of connection to lie on the line, reducing into a latent point. To meet the latent point requirement of $m_1$ to be on the line joining the two points, let us consider that the latent point is in the neighborhood, substitute its connection with $m_2$ as a latent material point so that, without loss of generality, the force $\boldsymbol{F}$ can act on $m_1$. The fourth point of connection in motion has no constraint; so its location depends on the motions of particles $m_1$.

Thus, the action and reaction forces have three independent points of connection; two points are material points and one is a geometrical in the direction of motion, and fourth latent point lies in the neighborhood on the line joining the two material points. The same way from Equation (1.4.1) it follows that the left hand side has three independent points of connection, and not four as noted before. In this case, one of the points is a latent material point associated with $m_1$ and lies in the neighborhood along the acceleration.

The three points connected force on $m_1$ leads to two cases; the first is along, and the second perpendicular to, the line joining two materials. The first case of the motion is well known in science as the Newton's falling of the apple. In this case the $m_1$ will move towards $m_2$ and the both will meet, and no further motion is possible. For the second case, the force $\boldsymbol{F}$ acts on $m_1$ along the line joining $m_1$ and $m_2$, but has initial velocity. Thus, the second case leads to two possibilities of Case 2a and Case 2b. The Case 2b is a motion of a body under the influence of a third body or two forces acting in the same plane. For the case 2a, the initial conditions are such that $m_1$ moves in a direction perpendicular to the line joining $MP_1$ and $MP_2$; $m_1$ will continue to move in that direction as discussed below.

**Case 2a** To illustrate these facts, let us consider $m_1$ to be the mass of the moon and $m_2$ to be the mass of the earth. Equation (1.4.1) describes the motion of the moon with $\boldsymbol{F}$ given by Equation (1.4.2), which describes a two dimensional elliptic motion. There is only a force $F$ along the line joining two points $MP_1$ and $MP_2$. There is no other force to produce any rotational motion in the moon; so there is no spin. The moon has a velocity along the tangent to the elliptic orbit, but no force along with it. The moon will continue to move along the elliptic orbit. The moon has three points of connection in its orbit; two points are located at the moon and earth, the third point of connection is along the tangent to the motion. In the motion of the moon, there exists a fourth (dependent) point of connection that is a latent material point that lies along the line joining the two particles $m_1$ and $m_2$ and is in the neighborhood of $m_1$.

The action of $m_2$ on $m_1$ appears as a reaction and it is on a line joining two points $MP_1$ and $MP_2$. The action and reaction have three points of connection, with a fourth point of connection being latent. At this stage it is important to note that there exists a fourth point of connection. This point becomes obvious and observable, as it happens when there are more than two bodies (friends) in any activities, their latent connections evoke. We will discuss this case in the following section. But before we discuss the case we need to clarify that the latent point of connection introduces a limitation on the Newton's third law.

From Newton's third law, a third point of connection for $m_1$ appears to be falling on a line originating from $MP_1$ that extends to $MP_2$. In the motion of $m_1$, it has three points of connection on the line joining $MP_1$ and $MP_2$. The motion of $m_1$ is maintained by creation of a fourth point of connection in the space in addition to these three points. The action and reaction are along the same line—both the forces have three points of connection—and the fourth point of connection produces elliptic motion in $m_1$.

In the above example, the $m_1$ is moving under a reaction satisfying Newton's third law of motion with an action from $m_2$, a body away from $m_1$. Per Newton's third law, the action and reaction are equal and opposite and act along a line joining the two material points.

The action and reaction act along the line joining the two material points requires generalizing when they are not necessarily along a line joining two particles, as is the case for the motion of electrons. The same way, the motion of a particle can occur under the influence of other particles, having more than two-material points of connection, as is the case of the earth under the gravitational effect from the sun and the moon. In this state of motions of bodies, the gravitational forces are not necessarily on the same lines and the limitations of particle theory in applying to the study the motions of celestial bodies become evident. In this case, the connections of two bodies—earth and sun, earth and moon—are on a plane but not on a same line, and cannot be reduced to set of two points; in this situation the relative distance between moon and sun is large and their gravitational interaction in the study of three bodies is neglected. For cases like these, let us introduce a postulate generalizing Newton's third law:

> *Action of material $m_1$ on a material $m_2$ with three-material points of connection in the neighborhood of a line in space S produces equal and opposite forces at each end point. These forces produce a motion with a fourth spatial point of connection in $m_2$, different from the other three.*

Under the above postulate, Newton's third law turns into a special case of two particles (bodies) in which the action and reaction are equal and opposite and are along the line joining the two. But, when there are more than three points of connections in the motion of a body, Newton's third law of motion needs to be generalized as noted above. The forces act in a plane, but they are not on the same line joining two material points, which we named as the Case 2b, and discussed in the following section.

## 1.5 Earth's motion under the influence of gravitation from the Sun and the Moon

Two bodies (friends) A and B deal in their communication and motion of acceleration with forces successfully with the required number of points of connection, and kept latent points if required. But when there are more than two bodies (friends), let us say A, B, and C, there is a good possibility that the latent points (facts), hidden differences, appear imposing additional (issues) motion. One can see these types of motions in amusement park rides. Particularly the spinning teacup-rides in which the teacups spin on a rotating platform, while the occupant(s) face each other or face the middle post, but do not spin, do not revolve. Or the Sky-Diver rides in which the Sky-Diver cars are mounted on a circular frame—like a Ferris wheel—that spins on a front-back axis similar to the barrel roll. One can see these facts in some YouTube videos that are available on many web sites.

Similarly, a practical natural example of these types of connections appear in motions of three bodies—moon, earth, and sun—moving under the gravitational forces of interaction appearing between two bodies as noted in the last section. In this case, the middle body—earth—has two separate points of connections, while the edge body—moon—has the latent point of connection.

The earth's motion is under the influence of gravitation from two celestial particles (bodies), the sun and the moon. For this motion, it is a three body (particle) problem with respective masses that of moon-$m_1$, earth-$m_2$ and sun-$m_3$, and their respective material-particle locations are at $MP_1$, $MP_2$ and $MP_3$. These three points form a plane, let us say Q, in the space S.

To simplify the discussion, let us consider the moon, earth and sun are perfect spheres and their mass distributions are uniform. Let O be the origin of the coordinate system CS with reference frame $K_1$ that lie at point $MP_2$ with the coordinate axes $OX_2$ and $OX_3$, respectively, along the line joining points $MP_2$ and $MP_3$, and perpendicular to the plane Q. Let us select an axis $OX_1$ perpendicular to $OX_2$ and in the plane Q. These three axes are fixed with respect to the plane Q.

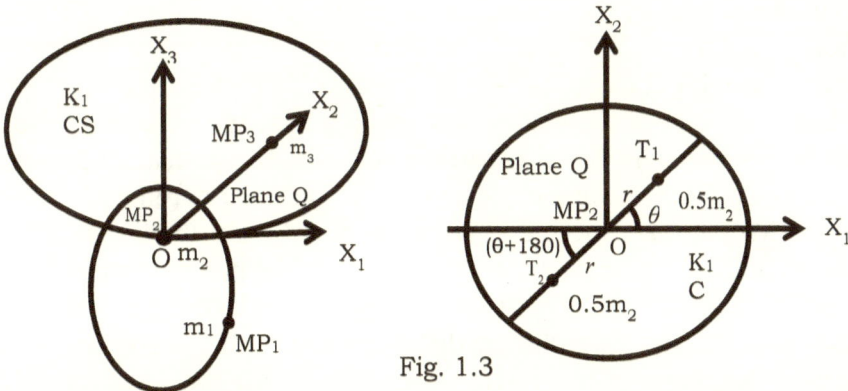

Fig. 1.3

The sun's gravitational field is stronger than that of the moon. As discussed in Section 1.4 above, the earth's motion under the influence of gravitational field of the sun satisfies the Equation (1.4.1), where the forces are given by Equation (1.4.2) with masses $m_2$ and $m_3$, and creates four-points of connection. The action and reaction of the sun and earth are equal and opposite, along the line joining MP2 and MP3 and remain so in the motion of the earth. All the four points of connection lie on the plane Q. The two points are MP2 and MP3 and the third, let us say $T_2$ is in the neighborhood of point MP2 and on the line joining MP2 and MP3. The $T_2$ is latent and can be relocated on the earth as long as the action of sun and the reaction of earth are along the line joining MP2 and MP3. The fourth point of connection, due to the moon, let us say $T_1$ is also free from an external force on the earth. The two points, $T_1$ and $T_2$, are on the earth and on the plane Q.

Under the gravitational field of the moon and earth, the action and reaction are equal and opposite, and the motion of the moon is the same as discussed in Section 1.4 above. But, the motion of the earth under the gravitational fields from the sun and moon is different.

For the earth's motion, the action and reaction from the sun will not alter the motion of MP2 and should remain in elliptic orbit, whether the moon is there or not in the space. The points of connection, $T_1$ and $T_2$, are free, connected with the point MP2 on the earth, and latent due to the other two particles, sun and moon, in the space and are associated with the motion of MP2. Under the

23

gravitational field of the moon, out of the three material points ($MP_2$, $T_1$ and $T_2$) of $m_2$, only $T_1$ and $T_2$ are permitted to change the motion of $m_2$ while keeping the motion $MP_2$ on the earth's elliptic orbit.

For simplicity in calculation and keeping the $MP_2$'s elliptic motion unaltered, let us consider two particles of mass $0.5m_2$ that are located on the plane Q on a diameter at a non-zero distance r from the center of mass $MP_2$ and are located at points $T_1$ and $T_2$. Let the line $(T_1, T_2)$ make an angle $\theta$ to the axis $OX_1$ giving rise to the polar coordinates to the points $T_1(r, \theta)$ and $T_2[r, (180+\theta)]$ as shown in the Fig 1.3. The selection of $T_1$ and $T_2$ and their distance r from $MP_2$ is so chosen to maintain the center of mass $MP_2$ moving under the gravitational field of the sun, and to create an entity with three-material-points of connection moving under the gravitational field of the moon.

Per the extended postulate of Newton's third law, the particles located at $T_1$ and $T_2$ will experience reaction forces from the moon's gravitational field and their magnitudes are given by

$$F = |F_{12}| = |F_{21}| = 0.5 K \frac{m_1 m_2}{r_{12}^2} \qquad (1.5.1)$$

where $r_{12}$ is the distance between the earth's and the moon's center of masses. The two forces at $T_1$ and $T_2$ are equal and opposite, and are at 2r distance apart producing a couple to rotate the earth around its axis and the acceleration, it follows that the earth spins, and is given by the equations

$$-m_2 r \dot{\theta}^2 = F \qquad (1.5.2)$$

In the derivation of equation (1.5.2) we have taken into account that r is fixed and there is no force in the transversal direction of motion. Per Equation (1.5.2), the earth has a constant spin, as $\ddot{\theta}$ is zero. The earth's spin can be expressed as a fourth point of connection located on axis $OX_3$. Let us denote the point of connection as $T_3$. The two forces F, each having three points of connection, act at a distance 2r, produce a couple giving rise to a six point connection. However, the couple has two constraints: first, the forces are equal, and second, they are parallel. These constraints reduce the six points spatial connection to a four point spatial connection for the earth.

The earth's mass $m_2$ has a total of three-material-points of connection and one spatial connection in the space, and moves under the gravitational influence from the sun and the moon. These four points are: the center of mass of earth, denoted by $MP_2$; the reactions from the sun producing elliptic motion of the earth with connection of having two points, denoted by $T_1$ and $T_2$ and are at equal distance from $MP_2$; and the reaction from the moon on the earth producing a spin of the earth with the forces acting at the points $T_1$ and $T_2$, denoted as a point $T_3$ along the spin axis $OX_3$.

Based on the day and night observations on the earth, a single fact raises doubts against the well-established particle theory; while using Newton's particle-point concept associated with the motion of celestial bodies, explains the elliptic orbits, but not its spin. After recognizing that there exists a limitation in the classical theory, and if we think it's business as usual to accept it, then we are living on island and doing disservice to science.

By extension of the concept of particle-point through connection resolves the limitations appearing when there are two set of gravitational fields acting on a body—earth. The larger field still acts and follows the Newton's second law, permitting to have elliptic orbit. Under these conditions, the earth and moon have the elliptic orbits. The weaker field demands to modify the Newton's third law and the cause for it is reasonable as the actions and reactions are not on a particle-point (the end point of the action and reaction of the forces), but lies on a line (which is perpendicular and is formed by two points at equal distance from the end point of the action and reaction line) as presented in the generalized postulate of third law of motion.

## 1.6   Comments on two types of forces originated by Gravitating particles and bodies producing elliptic orbital and spin motions

I wonder how people of the past, including the most revered, reputable, and respectable in science, have ignored and not even mentioned the phenomena observed, have not considered it unconscionable, and have never addressed it in their studies at all. This is the observed fact of the spinning of the earth and other planets which you and I see every day in our lives How could the founding fathers, so sublimed and devoted to the advancement of different motions have lived with—some participated in explaining the experiments to understand the various motion of bodies and that of lights—to maintain the Galilean Principle of Relativity (GPR). Or four hundred years latter how could the saintly Einstein, an implacable opponent of the wave theory and the reported limitations in Maxwell's theory for not meeting the GPR, nevertheless have never spoken or mentioned once in understanding the restrictions on forces appearing in the Newton's laws of motion?

In retrospect in judgement of explaining the spin, I do not feel superior to my ancestors; it really evokes the humility and respect for my teachers to have many open discussions and raised questions to look into the reality related to classical mechanics. Surely, some contemporary practices are looking into the highly expensive experiments to understand matter and its constitutions, rather than proposing some simple low cost solutions did not go with the discussions. The question is: Where are the simple solutions to the complex problems?

I (and many others, too) have long thought about understanding the material appearing in Newtonian mechanics, in Maxwell electrodynamics, and so on to be considered as particles or waves. In these studies, I am convinced that my readers will find it difficult to believe that we actually understood the motion of planets, and other bodies, but we did not look into the question why do they spin, irrespective of the material as particle or wave. Who cares what types of material they are made of? We need to understand why they spin.

The understanding of spin of the planets and their geometrical representation have been a salutary turn in our attitude towards the forces appearing in the third law of Newton, especially the one originated by the gravitational particles (bodies) and displayed in their motions.

The overriding mission of this chapter is to relinquish the restriction on the force of attraction $F$ between two particles (bodies), which has a maximum of four points of connection, and originates the action and reaction working along the line joining the two locations of the particles. The force does not directly participate in the motion of one with respect to the other. If the force $F$ participates in the motion, after some time, the motion of one will end as is the case of a falling apple. But in general, for large celestial bodies, it does not. But for these two cases, the force $F$ has three points of connection; the fourth one is latent. When a third particle (body) is introduced in the motion of the one, all four points of connection of $F$ appear to produce two types of motions in the body that come from the other two particles (bodies). Again the two forces of attractions from the other two celestial bodies do not participate in the motion of the one, but they produce elliptic orbital, and spin motions. For the gravitational particle having influence from two other particles will have two different forces of attraction to act on it; and one of the $F$ has three and the other has maximum of four points of connection, and no more. These facts are independent of the types of particles, bodies, or waves from which the forces originate. And, it is the four points of connection associated with the matter in the space-time manifold that easily explain why the planets with moons spin, but not their moons.

## 1.7 Summary and Conclusions

A particle at rest or in uniform motion has two points of spatial connection—one of the connection points is at the particle and the second is a geometrical point in space. In the case of the particle at rest that has two points of connection, the second point of connection is a latent point in the space.

A particle under the action of a force and its reaction appearing as acceleration has three points of connection. In this connection, the two points are two material points, while the third point of connection is along which the material moves. The force has three points of connection. In this case, the acceleration and also the force have a fourth latent point of connection which lies on a line joining the two material points, which does not participate in motion. The simplest possibility for a material point under the action of a second material point is an elliptical motion and the third point of connection lies on a straight line joining the two material points, and the Newton's third law demonstrates it. Under this condition, the particle cannot spin.

In general, the action and reactions are equal but not necessarily on a straight line. As an example, the forces in an electron's motion are not in a straight line. This introduces a postulate generalizing the action and reaction on the particles with three material points of connections.

A particle (body) under the action of a force in a plane of three-material-points of connection under two gravitating particles (bodies) is reacted by a spin motion. The spin originates a fourth-point of connection in the space, which is outside of the plane.

All gravitating particles-bodies have the maximum of four geometrical points of connection. These geometrical points are not required to be in the neighborhood. Similarly, for a three dimensional space, a coordinate system with a maximum of four points of connection—an origin and three coordinates—can represent the entire space. Due to these facts, it is possible to study the motion of a particle, a body, in space under a gravitational filed by considering space and matter absolute, the time remains absolute for both the items. Thus space, time, and matter are absolute for Newtonian Mechanics.

In the following chapters we will see that electrons, on the other hand, are particles with a spin and so must have a minimum of four points of geometrical connections, like a coordinate system. This is an inherent property of an electron. Thus, electrons appear as particles for Coulomb's law, but have a spin and so can also appear as volume, surface and waves too.

# Chapter II

# Union of the Space and Matter in Magnetic Fields

*The motions of Electrons
Unite the Space and Matter*

## Why?

**Because electrons have a minimum of four points of geometrical connections at rest and their motions that produce electromagnetic and magnetic fields require accommodating extra points of connections that unite the space and matter**

## 2.1    Introduction

Magnetic materials, called lodestones, were known to the ancient Greeks and others for a long time, way before the studies of motions of celestial bodies, electrical charges, and magnets had been conducted. After Newton, Hans Christian Oersted (1777-1851), Alessandro Voltas (1745-1827), André-Marie Ampère (1775-1836), Faraday, and others studied the properties of magnetic materials, electrical charges, and electrodynamics in about hundred years in the late 18th and 19th centuries. These developments remained under the direct influence of Newtonian mechanics, with the concepts of forces among two magnets or two electrical charges to follow the Law of Inverse Square of Distance between two particles, keeping the two entities—electrical charges and magnetic materials—independent. Based on Faraday's experiments, Maxwell brought these two independent entities under one roof and developed the equations of motions for electrodynamics.

Though the forces between two charged particles or two magnets follow the Law of Inverse Square of the Distance between two particles–two points, they are not of the same status. The forces in electrodynamics are relatively about billion-billion-billion times stronger than that of gravitation. Similarly, the forces are grouped together and derived from the actions between two particles, but the particles and their associated matter have essential differences. For electrodynamics, there are two kinds of matter which we associate with positive and negative charges, or with two kinds of magnetic poles—north and south; forces between like kinds repel, and unlike kinds attract. While for the gravitational forces, there is only one kind of matter always having attraction. Why do we have the material particles behaving so drastically different? What are the fundamental differences among the matter associated with these particles?

These material particles behave drastically different in their participation in the theoretical representations of motions in Newtonian mechanics and electrodynamics, as they are having fundamental differences in their associations—their connections—with the space. In the former case, the particles are large, visible, but non-associating and independent of the space; whereas in the later, they are very small, practically invisible but undeniably unite with the space (or mandate space to unite with time which we will see in Chapter III), that give them full privileges and supports to turn and twist in their fields and their motions without limitations of physical characteristics of the space.

In Newtonian mechanics, a state of motion of particles described by their gravitational fields depends on their relative positions and not on their velocities; the forces among these particles act at a distance without any express consideration of its surrounding. In Maxwell electrodynamics, a state of motion of particles described by electric, magnetic, and electromagnetic fields depend on the relative positions and their velocities; the forces among these particles and their fields depend on their surrounding neighborhood. In both the studies, the spatial-connections of matter is defined in Chapter I as *"a functional relationship of matter with a number of geometrical points in the space that participate in describing its motion,"* connects the matter with other particles in the neighborhood, or with the space, or both in which it is residing; the time facilitates to study and remains absolute in the study of their motions. So, we will focus on the union of matter with the space in this Chapter, and the union of space and time will follow as one of the benefits of the concept of connections without referring to the Galilean Principle of Relativity (GPR) that was required for the study of the Special Theory of Relativity.

The gravitational particles have two, three, or four points of spatial connections depending on the influence from the other particles in the space. These gravitational particles connect with themselves, but they neither unite with the space nor with the time.

The charged particles, referred here also as electrons, on the other hand, have a minimum of four-points of spatial-connections, and will be considered to have macroscopic (though invisible and has microscopic) structure in space with specific orientation. These four points of connections consists of: first, where the charge of material is located; second, along the electric field; third, along the magnetic field; and fourth, along its motion. Some of the points of connection may be latent depending upon how its activities are observed. Similarly, points in space have the minimum of four points of connections. These four points of connections consist of: first, where the point is located, while the other three points are along the three directions of the three dimensional space. Some of the points may be latent depending upon how its activities are observed and studied for the space, like the geometrical point on a line has a minimum of two points of connection, the surface has three points of connection, and so on.

The study of interaction of two electrons of four points of spatial connection appears as a special case, when the two charges are considered together, reduces to a three-point spatial connection, but the resultant force retains its orientation with the surrounding space. For this reason, in the static case, Coulomb's law, similar to Newton's 2nd law, produces two types of forces—the force of attraction and the force of repulsion depending upon the charges are of opposite or of the same orientation with the space.

In the static or the time independent electrons whose motions produce the electric, magnetic, or electromagnetic fields' cases, the electric and magnetic fields act independently of each other. Their interdependency appears when there are changes in the currents due to the motions of electrons or magnets around a conductor.

In addition to the above basic differences and limitations, there are formal and profound variations in the motions of these particles in space. In Newtonian mechanics, the action and reaction between two particles are equal and opposite and act along a straight line joining the particles and we can at the most discuss the motion of three particles. In general, this is not the case for electrons as they (mostly) follow the law of linear superposition.

In the case of electrodynamics, there are apparent asymmetries that are not necessarily intrinsic to the phenomena. For example, moving a magnet through a closed conductor at rest produces a current in the conductor and an electric field in the neighborhood of the conductor. But, moving a closed conductor through a magnet at rest, there is neither a current in the magnet nor an electric field in the neighborhood of the magnet; though there is an electric filed in the conductor.

From the above noted observations, two things follow. First, the steady-state motions of electrons produce electromagnetic and magnetic fields depending on how the electrons motion is connected with themselves and its neighboring space. These fields represent the spatial-connections associated with the matter and establish a union of space and matter. The time does not directly participate in the motions of electrons. The time on the other hand unites with the space, but does not explicitly demand to have the union and so one can consider time to be absolute in the study of Maxwell electrodynamics.

Second, under certain specific spatial (geometrical) constraints on the motions of electrons, the electromagnetic fields turn into magnetic fields which are independent of time. And there are certain materials like loadstones that have these magnetic field properties. Experimentally, an introduction of the material with magnetic field (magnetic pole) in a closed conductor induces the free electrons in the conductor to move, but its reverse is not true. So empirically, the electrons producing the magnetic field are in stable state of motions, and have higher number of points of spatial connection compared to that of the free electrons in a conductor (and electromagnetic field in space).

For a simplification in derivation of the union, in this chapter, we will consider the electrons to be of macroscopic structure moving with steady-state motions satisfying the Maxwell equations of electrodynamics.

The theory of union of space and matter to be developed is based on the electrostatic and magnetostatic fields produced by the steady-state motions of electrons—the kinematical association of matter of electrons with the space. It is the insufficient consideration of the union of space and matter that lies at the root cause of difficulties in the study of electrodynamics to divide the space into empty and occupied spaces.

## 2.2 Spatial-Connections of electrons at rest in space

Coulomb's law states that the force between two charged particles at rest is directly proportional to the product of charges and inversely proportional to the square of the distance between them. The law is restricted. The time dependent motions of charges violate the law. Under Coulomb's law the force between the two charges is along the straight line from one charge to the other. In this restricted case, as discussed in Chapter I on the spatial-connection, the charges do not spin, and both the electrons' and Coulomb's force have a three-point spatial connection that lies on a plane.

There are more than two charges present in the space and Coulomb's law is (and the Maxwell's equations are) supplemented with the "Principle of Superposition," which states the empirical facts of electrons:

> *The force on any charge at rest is the vector sum of Coulomb's forces from the other charges, and the resultant force is not necessarily along the line joining to any one of the charges.*

The principle of superposition leads to two succeeding results. First, the resultant force field is not limited to be along a straight line. It is along a curve. This is different from Newton's 3rd law of action and reactions that are equal and opposite and are along the straight line joining two points of the particles (bodies).

Second, the charges in the neighborhood produce an electric field that is proportional to the charges contained in the space. These facts extend Coulomb's law of electrostatic force field into a vector field with zero curl and a given divergence which, for simplification of geometrical representation, we associate them with the two electrostatic equations for the static case of the Maxwell electrodynamics. For the details of Maxwell's equations see Chapter IV. This leads to two cases for the spatial-connections for electrons.

**Case 2.1** George G. Stokes (1819-1903) developed a theory combining a line-integration of a vector field along a closed curve that can be expressed in terms of a surface-integration of its functional derivatives taken over the surface bounded by the curve. The vector field's line-integration over a closed curve is known as a circulation around the curve; its functional derivatives on the surface are expressed and are known as curl of the vector. This theoretically developed result is known as Stokes theorem.

The first case is that of a vector field whose curl is everywhere zero. Thus, according to the Stokes theorem, the circulation of the vector field around any closed curve is zero, as shown in the Figure 2.1A. In this case, if we choose any two points (1) and (2) on a closed curve in the space, the line integral of the tangential components of the vector field $E$, let's say an electrical field satisfying Coulomb's force field, from point (1) to point (2) is independent of a path taken. The electric field $E$ depends only on its initial and final stages, and can be derived from a gradient of a potential function, let us say $\varphi$.

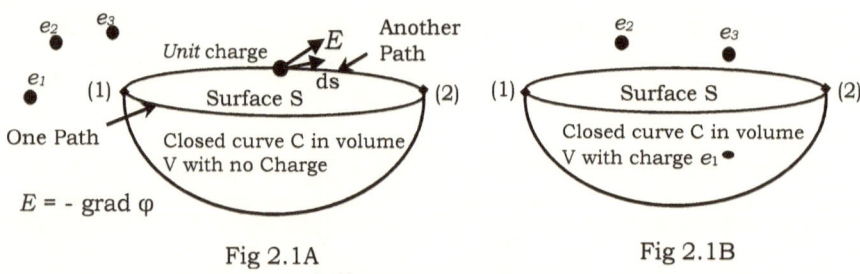

Fig 2.1A                    Fig 2.1B

From the above noted results it follows that for a vector field satisfying Coulomb's law, the charges and its associated field lines at rest are not (necessarily) straight lines, but are curves. Let us consider charges on a surface and their Coulomb's force field expressed as a potential function $\varphi$. Since each charge produces a field on a surface that can be locally expressed as a function of three points and so is for the potential function $\varphi$, each charge has a spatial connection with minimum of three points of spatial connections. Each point of connection lies in the neighborhood of the charge as the field depends upon its surrounding (neighboring) medium, and all of three points of connection lie on a surface. This is a fact and one can experimentally verify by charging a spherical conductor. In this case, charges lie on the spherical surface with three point spatial connections, there is no electrical flux through the surface; the electrical field inside the sphere is zero as there is no flux through the space enclosed by the sphere. Due to this, all charges lie on the surface not inside the surface. A fourth point of connection associated with the electrical charges exists when a closed surface encloses the charges, which we will see in the following Case 2.2.

**Case 2.2** Before Stokes, Carl Friedrich Gauss (1777-1855) developed a theory combining a surface-integration of a vector field along a closed surface that can be expressed in terms of a volume-integration of its functional derivatives taken over the volume bounded by the surface. The vector field's surface-integration over a closed surface is known as a flux from the surface; its functional derivatives from the surface are expressed and are known as divergence of the vector. This theoretically developed result is known as Gauss theorem.

The second case is that of a vector field whose divergence is everywhere zero. From Gauss' theorem on vector algebra it follows that for a divergence of a vector field from a volume V with a closed surface S enclosing finite charges, let us say $e_i$ (i = 1, 2, 3..), the flux from the closed surface is proportional to the charges contained in the volume. This case turns into two subcases of having no-charges and charge contained in volume V. Figures 2.1A and 2.1B geometrically represent these two subcases, respectively.

The first subcase is a closed surface S that does not contains a charge, and then due to the Case 2.1 above, the curl of a gradient of the potential over the surface is zero. Let us take any closed curve C in surface S with volume V as shown in Figure 2.1A with no charge; the curve C can reduce to a point on the surface S. In this case the electrical field can be derived from the potential function φ. By application of the Gauss's theorem, the flux through the closed surface S is zero, so the divergence of the electrical field *E* is zero. Thus, the divergence of the potential function φ is zero. In this subcase, there is no charge enclosed by the surface S with volume V, and the added point of connection for field $E$ lies at the charges. But each charge has its own geometrical point where it is located and three points of spatial connections are at the field $E$ on surface S. And, thus the charges $e_1$, $e_2$...have four points of spatial connections at each charge.

Let us consider a second subcase with closed surface S enclosing a charge with volume V. Let us say $e_1$, as shown in the Figure 2.1B. The result discussed is true for any number of charges. We will limit our discussion to a single charge $e_1$. In this case we cannot shrink a closed curve C enclosing the charge $e_1$ to a point. From the application of Gauss' theorem, the electric field $E$, produced by $e_1$, has non-zero divergence for the volume V and non-zero flux through closed surface S with volume V. To maintain the non-zero properties for the volume V and divergence of the field $E$—a change in the electric field with respect to the change in all direction of the space— the charge $e_1$ has to have in general a non-planer four-point spatial connection. The same way, for non-zero flux of the charge $e_1$ through a closed surface S with volume V has non-zero field $E$, a change in the electric field in all directions over the surface S has to have non-planer fourth point of spatial connection. In this subcase the four-point spatial connection is necessary and sufficient, respectively, to have non-zero divergence and flux through a volume V enclosed by a closed surface S for a charge $e_1$.

## 2.3    Union of Space and matter for electrons

The motions of electrons produce electromagnetic and magnetic fields in space. An electromagnetic field is due to an induction of free electrons' motion in space. The magnetic field is a resultant representation of motions of two or more unpaired electrons moving in a closed atomic orbit of a material, or in a closed loop of a conductor. The magnetic field is independent of time and also of the electrical field.

Faraday's concept of spatial continuity of magnetic field-lines leads to a local union of the matter of electrons with the neighboring space. The continuous magnetic field-lines are limited cases of three dimensional field curves on two dimensional surfaces in the neighborhood. The four point connected electrons (which Faraday, Maxwell, and others considered as bodies, so let us consider them to be macroscopic) in motion unite with points in neighboring space of each having a four-point connection and produce the electromagnetic and magnetic fields. To simplify our presentation, we focus on a steady-state motion of electrons. In this limited case, the union of the matter of electrons and the space, both having four point connection at rest, appearing in the electromagnetic and magnetic fields has, as noted before, eight point spatial-connections in its neighborhood.

Maxwell's concept of spatial continuity in the electromagnetic field due to the electrons' motions in electrodynamics extends the local union to a global union of space and matter through the equations of motion, in particular, through the divergence of the magnetic and electromagnetic fields to be zero, representing their changes in three dimensional space. These facts unite the space and matter globally.

In the time-independent case—in the steady-state motions of electrons—Maxwell's four equations (see Chapter V) turn into a pair of two equations representing electrostatic and magnetostatic equations. In this static case, the time does not participate in these equations, and electrostatic and magnetostatic phenomena appear independent of each other. For this reason, it is permissible and appropriate to consider the time to be absolute (Newtonian) in the study of static electromagnetism.

To observe the union of space and matter and to uphold it, we consider two well-known experiments of a steady direct current flowing through a closed (small circular) conductor and a straight wire. For the closed conductor, the magnetic field appears on the top and bottom (or the front and back) inside neighborhoods of the closed conductor. And for outside of the conductor and the straight wire, only the electromagnetic field appears in its neighborhood; for this case, we will discuss the case of the wire only.

The first case is analogous to the motion of unpaired electrons moving in a closed orbit of some material, influencing each other to produce a magnetic field. For simplification of discussion, we will consider the case of two unpaired electrons. The discussion equally applies to more than two unpaired electrons.

The second case is the motion of electrons moving on a straight conductor (wire) producing an electromagnetic field in its neighborhood. In this case there are two types of electrons—the first one are the real electrons moving on the wire; the second type of electrons are fictitious, induced in the space, appearing with the electromagnetic field produced by the motion of real electrons on the wire.

The four point spatial connected electrons in motion have velocity, and electric and electromagnetic fields, and each of them at the right angle to the other two. In the steady-state motion of electrons, the electrical field has its curl zero and a given divergence—expressed in terms of charge density which is constant in time; electromagnetic field has its divergence zero, giving rise to derive it from a curl of a vector—expressed in terms of current density which is also constant in time. The electrons with four point spatial connections can move on a closed circuit, or in a closed conductor and also on an open conductor—like a long wire. In either case electrons have to meet so as to maintain the forgoing required conditions for the electrostatic and magnetostatic fields to be independent of time, and satisfy the pair of equations. In both these cases, the time derivatives of the fields do not appear in the electrostatic and magnetostatic equations. In the following we will study these two cases.

**Case 2.3.1** Let $O$ be the origin of a suitable Cartesian coordinate system with three axes (OX, OY, OZ) as shown in Figure 2.2 and ($i$, $j$, $k$) unit vectors along these axes, respectively. We will first study the case of magnetic field produced by the motion of two unpaired electrons, denoted as $e$, in a closed orbit, located on opposite side of an atom or can be manually performed by producing a motion of sets of unpaired electrons located on opposite sides in a small, closed, circular conductor of radius $a$ as shown.

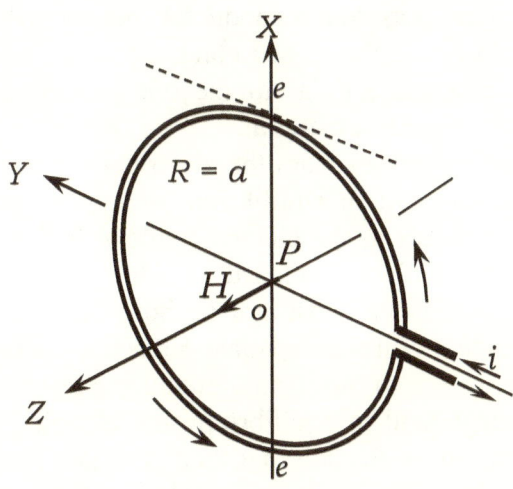

Fig. 2.2

The electrons moving in a closed conductor with charge density $\rho$ produce current $i$ and a magnetic field $H$ inside which is perpendicular to the surface of the closed conductor. The current $i$ and magnetic field $H$ are independent of time, are constant, and expressed in terms of the electrons. A simplification will result if two electrons, and not its time dependent motion—the current $i$—is chosen for the discussion related to the time independent magnetic field $H$. The final results presented here, without the loss of generality, remain the same whether we use $e$ or $i$ in our discussion.

Each of the unpaired electrons is moving in a closed conductor with a four point spatial-connection as shown in Fig. 2.2. Each of the electrons occupies a point on the conductor and produce electromagnetic field. We will see in Case 2.3.2, discussed below, that the electrons in steady state motion have eight points of spatial-connections in its neighborhood, six of which are independent. The set of two unpaired of electrons together will have twelve independent points of spatial-connection. Let us focus on the magnetic field produced by two unpaired electrons of a material moving in a closed orbit, or its equivalent in the closed conductor. Both the cases give the same results. For the conductor case, however, there will be an electromagnetic field outside the conductor, but we will focus here on the magnetic field inside the closed conductor and along the cylinder with conductor as edge, similar to the one produced in front of a magnetic pole of a magnetic material.

For a magnet, the forces from the electrical fields associated with the moving electrons are in equilibrium and their resultant electric field is zero. This occurs as the electric fields of both the electrons lie in a local plane-surface enclosed inside the conductor. This requirement originates one constraint on both the electrons and their available independent points of spatial-connections. The constraint on the two electrons, so as to move on a conductor and to maintain their associated electrical field in equilibrium, the electric field associated with each electron must be such that their resultant field lines joining them are along the radial-axis; the resultant of the electrical force fields is zero. This physical fact reduces two independent points of spatial-connections for the magnetic field.

Now, for both the electrons to remain on the surface enclosed by the conductor (orbit) with two points of freedom of motion in producing a common magnetic field, adding a second constraint reduces the two additional independent points of spatial-connection. Thus, the two unpaired electrons producing the magnetic field will have total of eight points of spatial-connection in its neighborhood, with two constraints. This fact leads to the minimum of six independent points of connection for the magnetic field.

In the case of a material having more than two unpaired electrons in the orbits, it has corresponding points of spatial connections and associated with two constraints to reduce the total number of geometrical points of spatial connection for the associated magnetic field. In a static case, this magnetic field superimposes the geometrical points of connection reducing to the same minimum of six point of spatial-connection in its neighborhood. We will see in Chapter III that the magnetic field and electromagnetic fields have more than six geometrical points of connections.

The observed experimental facts are: magnetic fields have north and south poles, without a monopole in the classical electrodynamics; and in quantum mechanics, monopole was theoretically presented to the magnetic fields, but was never detected. So, for the present study, we consider that there are two poles associated with the magnetic field and are separated by the magnetic material.

These two sets of magnetic fields located perpendicular to the two magnetic poles (magnetic material) need to satisfy the requirement that their divergence to be zero that give rise to a vector field derived from a curl of a vector. Thus, the magnetic vector field, a curl of a vector, has two equal and opposite forces, working at a finite distance from each other. The equal and opposite forces, each having three points of connection, do not add any constraint on each of the magnetic fields located on either side of the magnetic material, as the original six-point spatial-connections have satisfied this requirement.

The six-point spatially-connected magnetic field appears as a set of two separate fields, but physically that has never observed. The set of two magnetic fields always remain connected. The magnetic field on each side of the material satisfies the requirement that its local divergence to be zero. A curl of a vector expresses the set of the magnetic fields. The curl of vector is a third vector. The curl vector field, in addition to curl property, has to satisfy the global divergence to be zero. These three dimensional global properties connect the two sets of the magnetic fields, located on the top and bottom (opposite sides) of the material, and introduce a geometrical curvature (turn) and torsion (twist) in its three dimensional spaces of the magnetic field. The magnetic fields either on both sides of the magnetic material or that of the conductor have locally equal and opposite force fields at a finite distance of the plane separating these two fields create a curvature (turn), and its associated curl vector is not a straight as an arrow but has torsion (twist) to meet its local and a global properties of having divergence zero. Both the vectors are not independent, but are connected; the connection of the vector field produces more than six points of spatial-connection for the magnetic field. We will further discuss these cases in Chapters III and IV.

Both the electrons in the closed orbit or its equivalent in a closed conductor are under the action of the Coulomb's force field, giving rise to two equal and opposite forces $F$. The force $F$ is given by:

$$F = \frac{1}{4\pi\epsilon_o} \cdot \frac{e \cdot e}{4a^2} j \qquad (2.3.1)$$

The forces are under equilibrium under the centripetal forces due to the motion of electrons, either moving in the orbit or moving in the conductor and, is given by $mv^2/a$, where v is the velocity of the electron with mass m and $a$ distance from the center for the two forces. Please note that $e.e$ in equation (2.3.1) denotes a simple multiplication of e with another e giving rise to $e^2$. From the equilibrium conditions, then we have

$$F = \frac{mv^2}{a} j \qquad (2.3.2)$$

From which we get the following equations for two electrons:

$$\frac{1}{4\pi\epsilon_o} \cdot \frac{e^2}{4a^2} j = \frac{mv^2}{a} j \qquad (2.3.3)$$

These two equal and opposite forces give rise to a fourth point of spatial-connection along the magnetic field $H$. The magnetic field $H$ is a cross product of two equal and opposite forces along $j$, $2a$ distance apart along the radius vector and perpendicular to the plane enclosed by the conductor. The magnetic field $H$ produces a local couple of two equal and opposite forces, and is given by

$$
\begin{aligned}
H &= \frac{1}{4\pi\,\epsilon_o} \cdot \frac{e^2}{4a^2} j \times 2a i \\
&= \frac{1}{2\pi\epsilon_o} \cdot \frac{e^2}{4a} k \qquad (2.3.4)\\
&= \frac{\mu_o}{2\pi} \cdot \frac{e^2}{a} k
\end{aligned}
$$

The magnetic field is at the center, directed perpendicular to the plane, located on both sides of the plane. The magnetic field $H$ produced by two electrons is given by

$$H = \frac{1}{2\pi\epsilon_o} \cdot \frac{e^2}{4a}k = \frac{\mu_o}{2\pi} \cdot \frac{e^2}{a}\,k \qquad (2.3.5)$$

On the right hand side of the equation (2.3.5), the quantities $e$, and $\mu_0$, are magnetic field constants and $\pi$, a mathematical constant, do not change. So, from (2.3.5) it follows that the magnitude of the magnetic couple $H$ increases its value inversely as the value of $a$ decreases, the radius of the conductor or the radius of the orbit decreases without limit. In a limiting case, the magnitude of $H$ will go to infinity as the value of $a$ goes to zero, which makes no physical sense and it is not feasible. Similarly, the distance between two electrons cannot be reduced to zero to produce the infinite magnetic field $H$. From these comments follows the postulate:

*Postulate*

> *For unpaired electrons in motion in a closed conductor, or in a closed orbit of a material with radius R, producing magnetic and electromagnetic fields H in its neighborhood is such that the product of H and R is constant. The constant J, defined as HR = J, upholds the union of space and matter.*

From equation (2.3.3) and the kinetic energy of two electrons $2*[(\frac{1}{2})mv^2]$, we see that the constant $J$ has dimensions of $ML^3/T^2$, where M, L, and T, respectively, are the mass of electron, L the distance (Length) and T the absolute (Newtonian) time, and unites the matter and space.

Also, a magnet (just like gravitational body) acts as a body with eight points of connection, but has magnetic field (just like gravitational filed) to interact with other charges a closed conductor and induces current, which we will discuss in the Chapter III.

**Case 2.3.2:** As before, let $O$ be the origin of a suitable coordinate system with three perpendicular Cartesian coordinates (OX, OY, OZ) axes, and ($i, j, k$) the unit vectors along the axes respectively. A steady electron flow with charge density $\rho$ (rho) flowing through a straight conductor along the axis OX is shown in the Figure 2.3.

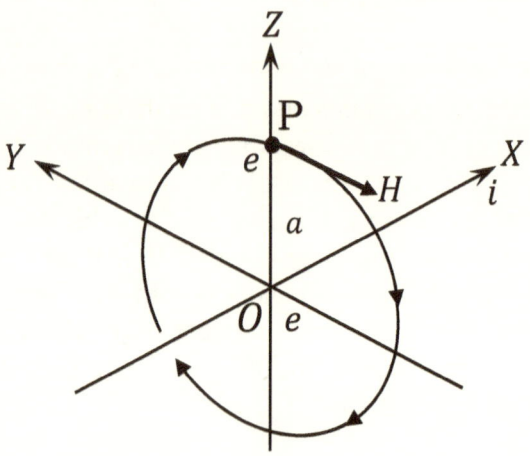

Fig 2.3

We focus our discussion on a single electron $e$ located at point O moving with a constant velocity $v$ on a conductor. The steady-state motion of the electron(s) in the long wire produces electric field $E$ and electromagnetic field $H$ in its neighborhood. We will focus on the static electromagnetic field $H$. The electron $e$ at O has constant mass m, velocity $v$ and $\frac{1}{2} mv^2$ kinetic energy, where $v^2$ denotes the square magnitude of the vector velocity $v$. As discussed in the Section 2, for the electron on the wire, the electric field $E$ has its curl zero giving rise to an electrical potential $\varphi$ (Phi). The electric field lines start and end on point O giving rise to a closed curve (surface) connection for the electron at O, establishing a minimum of three point surface, let us say S, for the electron in the space. The velocity vector originates from O is perpendicular to the surface S, establishing the fourth point of spatial-connection for electron at O.

Since the motion of the electron $e$ produces an electromagnetic field $H$ in its neighborhood, let us say at point P with a fictitious electron. The zero value of divergence of electromagnetic field gives rise to circulation around the curve enclosing the surface S which, according to the magnetostatic equation is equal to the flux of electric current density. From these facts, the electromagnetic field $H$ is given by:

$$H = \frac{1}{2\pi\epsilon_o} \cdot \frac{e^2}{4a} k = \frac{\mu_o}{2\pi} \cdot \frac{e^2}{a} k \qquad (2.3.6)$$

The surface S encloses the electromagnetic field and its direction is given by the right hand screw rule. It has three points of connection with a fictitious electron $e$ at point P. At point P, there is also an electrical field $E$ at right angle to the electromagnetic field which starts from and terminates at point O giving rise to a fourth point of connection with the fictitious electron at P. Point P is an arbitrary point at distance $R$ from O. Thus motion of $e$ at O gives rise to electromagnetic field $H$ with eight-point spatial-connections. The electrostatic energy due to the charge $e$ at point O and the fictitious charge at P with the electromagnetic field $H$ at a distance $R$ is given by:

$$\text{Electromagnetic energy} = \frac{1}{4\pi\epsilon_o} \cdot \frac{e^2}{4R}$$

The electron is in motion and its kinetic and electromagnetic energies are equal and can be expressed at a distance R as

$$\frac{1}{4\pi\epsilon_o} \cdot \frac{e^2}{4R} = \frac{1}{2} mv^2 \qquad (2.3.7)$$

The electromagnetic field $H$ in the neighborhood of the real electron at O and an induced electron at P has to satisfy that its divergence to be zero. The divergence constraint reduces two independent points from the spatial-connections. Thus, the electrons and the electromagnetic field *H* have six-point spatial-connections in its neighborhood.

For the electromagnetic field *H* at distance *R*, by following the same reasoning of Case 2.3.1, the above noted postulate follows, and we have the result as *H*R = *J*, where *J* is a constant that upholds the union of space and matter.

## 2.4    Conclusion

1. Electrons at rest have the minimum of four points of spatial-connections.
2. The steady-state motion of electron produces electromagnetic field *H* in its neighborhood. The electron and its electromagnetic field *H* have six-points of spatial-connections.
3. The steady-state motion of unpaired electrons in a closed conductor or in a closed material orbit produces magnetic field *H* in its neighborhood which is perpendicular to the conductor or to the orbit. These unpaired electrons and the magnetic field *H* have six-points of spatial-connections.
4. The steady-state motion of electrons unites the space and matter through the magnetic and electromagnetic fields. The union of space and matter is represented as magnetic and electromagnetic fields *H*, produced at a distance R from the moving electron, and in its neighborhood satisfies the equality *H*R = *J*, where *J* is a constant.
5. Thus, electrons at rest have up to four-points, and in steady-state motion their matter unites with space with six-points of spatial-connections in the neighborhood.
6. A magnet acts like a body and its motion in and out of a closed conductor excites the electrons and introduces current, which will be discussed in chapter III.

# Chapter III

## In motion of electrons and magnets Time unites with Space

## Why?

Because electrons and magnets, respectively, have minimum of four and six points of geometrical connections and their motions require to have added points of connections with a constant velocity—which is equal to the velocity of light—that unites the space and time.

## 3.1    Introduction

The rise of accepting the constancy of velocity of light to maintain the Galilean Principle of Relativity (GPR) for the invariance of Maxwell's equations of electrodynamics first took scientists by excitement in the early twentieth century. But its consequences are baffling to many even today that how the velocity of light maintains its constancy when observed by two observers, two entities moving under the uniform relative velocity. Maintenance of constancy of velocity light is contradictory to our day to day experience of adding two velocities of two bodies or two particles, either going in the same directions, or in opposite directions, or perpendicular to each other; how it maintains one of the bodies—the velocity of light—invariant! And under the constant velocity of light: Why is there no fictitious force in the Maxwell's electrodynamics when considered under the non-uniform motion of the reference frame which is the case for the Newtonian mechanics? These questions need resolutions.

For most of the 20th century, relativists and quantum mechanists in physics and mathematics faced family feud encounters among themselves in their progress to face the battles: against considering electrons as particles; against accommodating the wave nature of electrons to satisfy the particle nature, similar to that of photons that follows from the quantum mechanics, having roots in Newtonian mechanics and in Coulomb's law of electrostatic. To accommodate particle nature and to compromise with the waves conditions faced by these encounters, it was easy to settle the issues by considering the matter with particle/wave duality, rather than looking for the root cause of the problem associated with the matter: Why and how does matter associate with the electrons acting differently as particles and waves in its motion with respect to the space and time?

The concept of connections addresses both the questions, and we will present the answers in this Chapter through the Ampere's and

Faraday's experiments. The legitimacy of the answers lies in these experiments.

For the bodies (particles) at rest and in motion in the Newtonian dynamics and Maxwell's electrodynamics, respectively, have similarities and dissimilarities which appear to be non-inherent to the phenomena, and the equations of motions reveal the mathematical constraints on the theories they represent. In the similarity of the bodies at rest, the force of attraction (and repulsion) between the two is proportional to the product of their masses or unlike (and like) charges, and inversely proportional to the square of the distance between the two. The bodies in motion have ample dissimilarity. For examples of motions in Newtonian dynamics—the force acting between two bodies depends on their masses and their relative position, but neither on their velocity nor on their surroundings, and the space and time are absolute; in Maxwell's electrodynamics—the force acting between two electrons depends not only on their charges and relative position, but also on their velocity and their surroundings, and the matter and space are no longer absolute, and we have demonstrated in Chapter II that they are united.

In the development of the theory of motions of electrons and in understanding of its dissimilarities with the Newtonian dynamics, the concepts of space by itself and the time by itself were tacitly removed through an indirect union of space and time. We will demonstrate that the union existed and will reveal that it had sprung from the experiments of Ampere, Faraday, and others, without explicitly recognizing it, but realized to use it, way before the derivation of Maxwell's equations and the foundation of the Special Theory of Relativity (STR).

The development of STR requires two postulates to maintain the invariance of the Maxwell's equations of electrodynamics under the uniform translation motion of reference frames. First: The light and electromagnetic waves propagate with a definite, constant velocity of light *C*, which is independent of the state of motion of the emitting matter or that of the observer. Second: The Maxwell's equations need to follow the Galilean Principle of Relativity (GPR). These two postulates are incompatible. The STR has helped to rid the concept of ether, introduced the four dimensional space-time where time appears as another dimension like the three dimensional space, and explained the physical phenomena of having an inconsistency in the Michelson and Morley experiments (1877–1887).

After the introduction of constancy of velocity of light, we have heard the noise either from the theoretical discussion on particles—like Tachyons—or from experimental observations of detecting material particles—like subatomic elementary particles—travelling faster than the velocity of light. A statement of material travelling faster than the velocity of light demands to have a clear understanding of the foundation of the union of space and time, which is crucial in the motions of electrons, electromagnetic and magnetic fields, light and other subatomic particles. These particles and the associated fields travel with much higher velocity compared to the one appearing in the Newtonian mechanics. So, in the union of space and time, or that of the space and matter, we need to be clear on the circumstances and conditions under which the formation of the unions occurred.

The time expresses itself by revealing its presence in the movement of matter in terms of its changes in the space. If the matter does not change in its structure or in its location with respect to a reference frame of the space, we cannot observe and report time. There is no time if there is no change or displacement of matter in space.

A continuous motion of matter—the hands of a clock, or the vibration of an atomic element—represents time as a movement of material relative to the neighboring space. The time—the absolute or the relativistic—by itself is lame, but operationally effective when observed in motion of celestial body, or united either with the space or with the matter that permits to observe the motion of matter in space. Thus, the time can be absolute, or unite with the space or with matter or with both. We will focus on the union of space and time.

There is a difference in the motion of particles appearing in Newtonian mechanics and Maxwell's electrodynamics. These two sets of particles, respectively, have maximum and minimum of four points of spatial connections. These facts raise a question: Why should there be an external additional restriction on the Maxwell's equations to maintain the invariance under the GPR? We do not need to impose the incompatible external restrictions of GPR and the constancy of velocity of light as the phenomena comply with both of them when studied under the union of space and time; the union permits to have non-uniform motions for the induction currents.

We will show that Ampere's laws on magnetic induction and Faraday's laws on induced current are adequate to establish the union. For the phenomena, the union reveals that the non-uniform motion associated with the reference frames (matter) produce real forces, not the fictitious one appearing in the Newtonian mechanics.

It is the two incompatible postulates noted above working independently in non-cooperative way permitted to have the noise for the material traveling faster than the velocity of light. The union of space and time established here under the spatial-connection of matter will mute the issue, remove the dependency on both the postulates, and will provide an alternate explanation for having the constant velocity of waves of electromagnetic fields including the photons.

## 3.2    Essentials of the unions

The motions of unpaired electrons in atomic orbits produce magnetic fields that do not explicitly depend on time leading to the union of space and matter, as discussed in Chapter II. In this union, the matter and space do not have the equal status. The matter plays a major role in the union. The matter unites with the space. The space unites with the matter but the union achieved as a function of two material particles at a distance (in the space); in this sense the space acts as subordinate and supporting player of the union. The union is a function of motion of two electrons. The electrons' motion associates with the time; the time remains tacitly inactive and does not explicitly participate in the union of space and matter.

The developed union of space and matter, and the proposed additional union—the union of space and time—are neither perpetual nor everlasting. They are local, associated with the matter, and are permitted to have motion of matter in space associating with time; these two unions permit the matter with different characteristics to have different permissible motions in the space and the time. In this sense, in this chapter, we will see in the Section 3.7 that the motion of matter occurs with additional points of spatial-connections in the space and time. These added points of spatial-connection are dormant (when studied under classical Maxwell's equations), but permit to maintain the consequences of motion and GPR. In Maxwell's electrodynamics, the time unites with the space; the space and time remain independent of the motion of the material, but are not absolute. These facts are contrary to Newtonian mechanics in which these three entities—the space, the time and the matter—are absolute, and one does not associate or connect with other two.

One cannot easily comprehend to foresee all possible motions from the theoretical equations of motion until the material happens to be in a situation to have a permissible occurrence of the motions at an opportune space and time. It is the freedom permitted by the union—either with the space or with the time or with the space-time—that permits the matter to have all possible motions in the space and time, producing a possible motion not contemplated by the equations of motion. We will focus on the union of space and time, and will refer it as space-time.

In these unions—space and time, and space and matter—the matter is the primary player. The space and time are secondary, permitting the matter to reside in it to have the motion. And the time permits the motion of matter of electrons or that of magnets in the space. The space and time accommodates the matter to have various possible motions. It is the matter that unites with the space or with the time or permits to have the space-time, while the space, the time, or the space-time remains indifferent to the matter for a specific union of space-time.

## 3.3   Setting of time

Let us consider a reference frame K in which the equations of Newtonian mechanics and Maxwell's electrodynamics are valid. For a description of motion, one can introduce a coordinate system; let's say the Cartesian coordinate system in the reference frame K. To have a clear understanding and to distinguish other reference frames, let us say K' that may not necessarily be at rest, we will consider the reference frame K to be stationary. The motion of material point takes place in K and expressed as a function of time. In this expression of motion, time appears as known and given. Is the time known and given?

In an application of time to describe motion of material has no physical meaning unless we are clear and understand: What is time? We have to be careful and clear in the mathematical representation and usage of time. There are three (observational) concepts associated with time. These concepts amend themselves to move forward in the development of physical theories.

First, the conventional concept: The time performs as a continuous entity in motion of a particle in space. For a motion, there are presumptions that the flow of time is uniform, continuous, unidirectional, and in the neighborhood of the motion—local. If any one of the presumptions violated, the time and motion are not well defined. For example, 86,400 seconds (or 24 hours) are well defined through the earth's spin around its own axis. This definition of a second measures time on the consistence, continuous spinning of the earth and duration between two consequence seconds is uniform, local and unidirectional. (Following the SI units, one can define a second through the period of vibration of the Cesium atom.)

The time moves—from the present to the past; the future arrives as the present. The time depends on the motion of a material in space—changes with events—with no origin (zero) of its own; it is only observable and reportable through the current events—occurring at the present. The past is history and the future will come for the motion of a material.

This definition of second follows from the earth's motion. We transfer the concept of the second to the motion of other materials provided that the motion of the material follows the same criteria of time, and thus associate with it that leads us to the second concept.

Second, the Newtonian concept: We measure time with occurrence of events—motions—that depend on the space and the material. For example, a train arrives at 10:00 AM at a specific platform, say at platform A, we mean the following: The arrival of the train at A and the small hand of my watch—working uniformly based on the definition of the second from the earth's spin—at 10:00 at location A are simultaneous events. This is the way we have transferred the time from one event to the other.

Here we have presumed that the train and the watch are in motion and none of them have stopped working—moving. If any one of them failed either in their motion or in their locations, then the events are not simultaneous. For example, if the watch motion is non-uniform or either I am not at A, or the arrival of train is other than the platform A, I am not sure about the arrival of the train and my watch's small hand reaching at 10:00, are the simultaneous events. These simultaneous events reveal their limitations permitting to introduce a restricted union of space and time for the Newtonian mechanics. The space and time are absolute and non-connected; so the union is restricted.

Due to non-connected space and time in Newtonian mechanics, the equations of motion of a particle have influence from other material particles creating real forces to act at a distance that transmit instantaneously. The space and time themselves remain absolute, and do not directly influence the motion of material particles, except introducing the fictitious forces. Neither the material, nor the space changes with the time defined above.

The Newtonian equations of motion remain invariant under two-types of changes—motions—relative to the frames K and K'. First: The frames K and K' can have a simple finite relative displacement in their position. In this displacement, the change is a geometrical one for the space only, and time does not participate. This is a geometrical displacement and the Newtonian mechanics disregards its effect. Second: The frames K and K' can have a relative uniform translation motion. The uniform translation motion, expressed as a constant relative velocity $V$ between two frames unites the space and time. As a zero point of time has no meaning, both the frames use (treat) the same time for the motion. If $V$ is not constant, it will add fictitious force in the equations violating the laws of Newtonian motion. The space and time maintain the invariance of the equations of motion and meet the Galilean Principle of Relativity (GPR) when the velocity $V$ is constant. This leads to the Newton's first law of motion—equating the rest with the uniform constant linear motion of materials and that of the reference frames. This fact about K and K' creates a union of space and time with a constant velocity, known as inertial frames. The inertial frames with a constant velocity $V$ having a finite duration of time permit to have two finite distance apart spatial points in space in one coincides but not in the other. This is not the case for electrons, and their motions and associated fields.

Third, the Maxwell's (and STR) concept: In the Maxwell's electrodynamics, electrons and electric, magnetic, and electromagnetic fields are present in the space even there is no physical charge; in their motions, they alter with the space and time. The space and time are no longer absolute, but united to influence the motion. The united space and time require that for a change in time or a change in space of the material and their fields cannot remain the same; they must change with changes in time in the space S during their motions. The changes in charges and fields expressed in terms of space and time appear as partial differential equations of motion in Maxwell's electrodynamics. While in the Newtonian mechanics the motion appears in terms of ordinary differential equations. This is a major and a fundamental difference between the Newton's and Maxwell's equations of motions.

Let us analyze the velocity $V$ for a set of electromagnetic events that moves from a point $P$ at time $t_1$ to its neighborhood to a point $Q$ at time $t_2$, where $P$ and $Q$ represent the position vectors with respect to the origin of the reference frame K. For these events, the difference of the times $(t_2-t_1)$—the duration D—have no origin of its own, and it is independent of the reference frame K, and will also be the case for K'. They depend only on the events and the points where recorded. All these entities directly associate with two points in the active motion of the events in the space S. Thus, we have two points with spatial distance between them; say $L$, and two different times with duration D. For the observed velocity $V$, these two entities are nonzero and independent of the origin of K and K'. The velocity $V$ related to the events depends only on the spatial distance $L$ and the duration of time D, and represented by Equation

$$V = L/\text{D} = (P - Q)/(t_2 - t_1) \qquad (3.3.1)$$

The $V$ unites space and time for the Maxwell's electrodynamics, where spatial distance $L$ cannot reduce to zero with nonzero time as it was the case for Newtonian mechanics. This physical reality is true for K and K', where K and K' being either at rest or moving with a relative velocity $v$ in its neighborhood.

In the above discussion, there is no restriction on the magnitude of $V$ and it represents a case of a motion of the electromagnetic events without associating $V$ with the velocity of light. This fact removes the improvised introduction of constant velocity of light of the STR.

Using Boolean algebra, by using algebraic logic, one can systematically and easily verify that the constancy of velocity of light appearing in STR does not deny of existing particles travelling faster than the velocity of light.

So, the results are applicable to any non-Newtonian fields, including Tachyon and other atomic particles, are conceivable to travel faster than the velocity of light, provided their events permit to have a duration of time that reduces to an infinitesimally small quantity and have a very large distance between the two associated spatial points with the corresponding times of events to occur. The important thing in the presumption of $V$ to be larger than the velocity of light is that the non-Newtonian fields and its associated particles need to exist meeting the stated requirements.

## 3.4    Ampere's Law of induction

Hans Christian Oersted (1777–1851) observed that a long wire carrying (steady state) electric current produces deflection in a suspended magnetic needle when placed in its neighborhood. The magnetic induction is in a circular pattern. The pattern indicates the magnetic field lines produced by a current are concentric circles with the center on the wire. Jean-Baptiste Biot (1774–1862) and Felix Savart (1791–1841) first and then in more details through experiments, Ampere established the basic force law relating to magnetic induction $H$ at a point Q in the space due to the current in the wire at a point $P_1$, or to the currents between one current to the other in two wires.

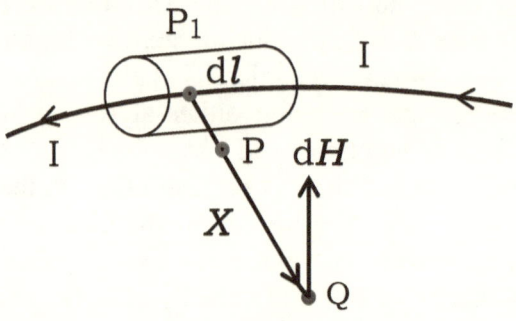

Figure 3.1

Let d*l* be an elemental length at $P_1$ in the direction of the steady flow of current I in wire and *X* be the position vector from the elemental length to the observation point Q as shown in the Figure 3.1. The elemental flux density of the magnetic induction d*H* at the point Q is given, in the direction and magnitude, by

$$\mathrm{d}\boldsymbol{H} = k'I \frac{(d\boldsymbol{l} \times \boldsymbol{X})}{|\boldsymbol{X}|^3} \qquad (3.4.1)$$

The symbol x denotes the cross product of two vectors. A word of caution about the Equation (3.4.1) at this stage is appropriate. The Equation (3.4.1) is meaningful only as one element of a sum over a continuous set, the sum representing the magnetic induction of a current loop or a circuit. Otherwise, it will violate the continuity equation of the divergence of the magnetic field that is always zero.

To meet the noted divergence of the magnetic field requirement and to overcome the difficulty associated with the local elemental current I, let us recall that the current I is actually charge in motion. We can replace I d*l* by q*v* where q is the charge, *v* is the spatial velocity of the charge, and the local magnetic induction d*H* by the magnetic induction *H* at point Q. For the union of space and time, we consider that the spatial velocity *v* is small compare to the velocity of the light *C*, and it is in a steady state motion. Thus, the flux density for such a charge in motion would be

$$H = \text{k}'\text{q} \, \frac{v \times X}{|X|^3} \qquad (3.4.2)$$

The constant k' appearing in the Equations (3.4.1) and (3.4.2) is a product of two constants: The first constant depends on the properties of the material through which flux is propagating. Let us denote it by $\in$ and note that its value lies between 0 and 1; for the ideal space its value is 1. The second constant is directly associated with the empirically observed velocity of light. We will temporarily express it as a constant k; its value is equal to the magnitude of velocity $|V|$ which, per our discussion of Section 3.3, establishes the union of space and time.

The flux $H$ is zero if the current or its equivalent the charge is at rest, or $\in$ is zero. The motion of electron with velocity $v$ produces two types of moving events in the space. And thus, the time appears at two places in Equation (3.4.2). First, the time appears in the velocity $v$ of electron that produces the current in the wire and it is absolute and Newtonian one.

If the velocity $v$ is high, the electrons will have high energy forcing it to move out of the steady state orbital motion from its (atomic) shell. The electrons will jump from lower shell to the higher shell. When the electron loses the added external energy, it falls back into the shell of its steady state orbital motion. The released energy appears as photons leading to the relation of Planck constant of quantum mechanics, which we exclude here from the further discussion.

Second, the time appears in the propagation of the magnetic induction $H$. The magnetic flux originates due to the motion of electrons producing current I that depend on the velocity $v$, and also on $V$ that depends on the events associated with space and time. To simplify our discussion, let us consider that the flux propels in space creating two events at two spatial points P and Q in the neighborhood, with their associated times. We have selected the point P near the point $P_1$ as a different point to simplify the geometrical representation of the discussion to follow, and to have similarity of presentation done in Section 3.3. The separation of the two events and associated duration of times to travel from the point P to the point Q have non–zero distance, non-zero time and non-zero velocity and the Equations (3.4.1) [and (3.4.2)] unite these three quantities.

The motion of electrons originates the events of the magnetic flux. The flux originates at $P_1$, propels through space, and flows from the point P at time $t_1$, reaches point Q at time $t_2$ and continues. The Equation (3.4.1) expresses the magnetic flux (propelling) velocity $V$ uniting the space and time.

## 3.5 Velocity of light (C) need to be constant for the magnetic flux

The magnetic flux appears not only at points P and Q, but also in its neighborhood forming a complete circle around the wire as its center. The flux propagates away from the wire sustaining the union of space and time.

In the propagation of the flux, the space, the time, and their union are continuous. The continuity of one is in harmony with the other two. A violation of any one's continuity is unlikely as three of them are interconnected.

To maintain the continuity of the velocity (of light) $V$, it must be constant in the space and the material through which it propagates. If it is not, than one can choose the distance between two events with the distance $L$ in space, and the duration in time D so that one of them is zero while other is not, which will violate the continuity either for the space or that for the time, which is not permitted. So, this requirement is satisfied only when $V$ is constant.

The flux will change when there is a change in the material occupied in the space. In this change, only the values of $\epsilon$ changes, but neither the magnitude nor the continuity of velocity (of light) $V$ will change. The $V$ remains constant. The constant magnitude of k' adjusts from one value to the other accommodating the material characteristic changes. Thus, the change in flux is a property of the material. The flux is neither changing nor influenced by the properties of the space or the time or the constancy of the velocity $V$. So, let us amend the Equation (3.4.2) as follows.

$$H = k\epsilon q \, \frac{v \times X}{|X|^3} \tag{3.5.1}$$

where $\epsilon$ denotes the material property and k is the magnitude of the velocity $|V|$ as discussed in Section 3.4. From Equations (3.4.1), (3.4.2), and (3.5.1) it follows that the k, the ratio of the electrostatic units to the electromagnetic units is empirically found to be equal to $1/C$, where $C$ is equal to the velocity $|V|$. This modification permits to maintain the union of the space and time, and defines the constant velocity of light which is given by $C$ =299,792,458 *m*/s. In this union, the derived $C$ is constant, and is independent of one of the prerequisite assumptions in the development of STR to meet GPR.

In the above definition of the velocity of light, we deliberately stayed away from speaking of the space to be empty or occupied with matter, as we consider that the matter occupies the space which is not empty. We introduced k as the ratio of the two set of units, known as Gaussian units. The ratios of the electrostatic units—defined for the charges at rest—to the electromagnetic units—defined for the charges in motion, introduces a connection between two spatial points with time. This is the property of the space and time for the motion of electrons that unites them with the velocity of light. For maintaining the union, the velocity of light must be constant.

## 3.6    Spatial connection of velocity of light

From the Equation (3.5.1) it follows that there are two velocities associated with the motion of electrons. One is the conventional velocity *v* which uses the (Newtonian) absolute time in its definition and has only two geometrical points of spatial-connection, and the time is absolute. The second velocity originated from the motion of electron is empirically equal to the velocity of light. This velocity $V$ has higher number of geometrical points of spatial-connection than that of *v*, the first one. From the Figure 3.2 it follows that the $V$ depends on its two points at P and Q, the associated times $t_1$ and $t_2$, and the electric and magnetic fields observed as two events at the two points due to motion of the electron at $P_1$. The points P and Q represent the geometrical points, and without the loss of physical and geometrical properties, we can consider them as geometrical points of the events.

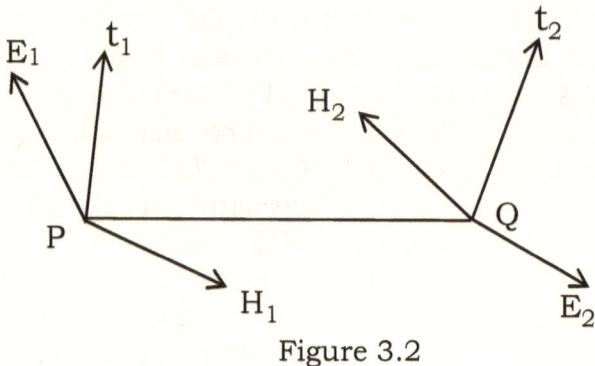

Figure 3.2

Thus, the velocity of light has eight points of spatial-connection: two are geometrical points, two for the two different times at the two locations—as the time is considered to be similar to the geometrical points in the union of space and time—two each along the electrical and magnetic fields.

## 3.7   Faraday's Law of Induced Current with the EMF Complying with GPR

Before Faraday's reasoning, observations of electric currents producing magnetic fields in space or in other wires suggested that (motion of) magnets may originate electric current in wire or in space. Neither a large magnet nor a substantially electromagnetic field due to the large current in a wire produced any electric current in a second wire or in space. Based on similar reasoning of Newtonian mechanics, it was tried without success, as there was no electric current or electric field in the second wire or space by non-uniform motions of electrons in a wire or in a magnet by considering it to be non-inertial frames connected with the electrons.

The above noted empirical observational attempts suggest that the phenomena of electric currents and magnets have a spatial-connection uniting the space and time, and an individual physical entity is not capable to produce a current either in the space or in a wire. The reason for inaction is that a current carrying wire or a magnet changes either with the time or with the space; but not with their union. So, the current carrying wire or the magnet is unable, incapable, and ineffective to introduce a current into a circuit or into other wire. Faraday's experiments demonstrated how he overcame these limitations by introducing three different possible motions producing the space-time changes in the circuits. By circuit we mean, as it is currently understood in the physics, a current carrying wire that encloses a geometrical surface in the three dimensional space.

Faraday discovered the first quantitative changes in electric and magnetic fields and demonstrated by experiments on behavior of currents in circuits. He reported a transient current induced in a circuit under one of the three possible changes in its neighborhood. (1) When adjacent circuit with a flow of steady state current is moved relative to the first circuit. (2) When one of the pair of wires has change in current, like the turning on and off of the current in one of the circuits, a current is induced in the other. (3) When a single pole of the permanent magnet is moved into or out of the circuit. This is an induction effect that induces a transient current attributable to a change in magnetic flux linked to the circuit. The changing flux induces an electric field—electric force—around the circuit is known as the electromotive force (EMF). The EMF causes a current to flow in a circuit. The EMF is defined as the tangential force per unit charge in the wire integrated over the length of a closed circuit. Faraday's experimentally observed discovery of the EMF demonstrates to generate an induced current in a circuit in three different ways: by moving the current carrying wire near the circuit; by changing the current in a nearby circuit; or by moving a magnet in and out of the circuit. Any one of these three changes originate EMF on the free electrons to flow in producing the induce current in a circuit.

Faraday's experiments related to three different motions reveal three physical facts about the motions of electrons producing induction currents in wires through the EMF associated with the circuits. These circuits are topologically (kind of geometrically) closed loops enclosing surfaces in the three dimensional space S. And the motion of electrons in two dimensional curves (circuit) establishes connection with the three dimensional surfaces enclose them.

**First Fact:** When a Current Carrying Circuit Wire (CCCW) with steady current moves relative to the second circuit wire, it originates a relative motion of the union of space and time associated with the second one. The steady current has eight points of spatial connection. The motion of CCCW adds additional spatial points of connection to the flow. There is no constraint on the motion of the first circuit, creating a three dimensional motion with three points in the space and one point associated with the time creating a total of twelve points of spatial space-time connection associated with the CCCW. The CCCW with the union of space and time has twelve points of space and time connections. In these twelve points of connection, the time appears twice in the connection, as it was the case in the Section 3.4. First as an absolute Newtonian time, and the second as the time united with the space. We will refer to these twelve points of connection also as the Spatial Space-Time (SST) connection. The twelve points connected SST connection of CCCW is forcing the four points connected free electrons in the second circuit (wire) to flow to produce an induction current with an eight points of spatial-connection.

**Second Fact:** When the current is flowing through one of a pair of wires, the free electrons in the wire flows with velocity $v$ out of its normal steady state motion causing resistance between atoms creating heat. The velocity $v$ is small so that electrons do not move out of their (atomic) shells as stated, and required in Section 3.4. The electrons, however, have eight points of spatial-connections in the motion. The change in the first wire with the current changes the velocity $v$ which adds additional points of connection to the motion of electrons. The change in velocity $v$ of the flow of electron is an additional change in its normal steady orbital motion around the nucleus. These two changes together are three dimensional and time dependent creating additional four points of connection into the existing eight points of connections. Thus, we have the twelve points of SST connection due to changes of the flow of current through one of the pairs of wires originating a motion in the four points connected electrons to flow in the second wire of the pair producing an induction current in it.

**Third Fact:** Each magnetic pole has eight points of spatial-connection uniting the space and matter of two or more electrons in orbital motion of atomic elements. When a magnetic pole is moved in or out of a circuit having a three dimensional closed curve, it is cutting a three dimensional surface S, adding three geometrical points and one temporal point to the eight point spatial-connection of the pole which originates a twelve points of SST connection. These twelve points SST connection of the magnetic pole again excites the four points connected electrons in the circuit causing the electrons to flow in the circuit originating an induction current in the circuit.

From the above presentation we see that the effects of the phenomena of the induction in space and in circuits, and their geometrically complicated representations are topologically interesting. Faraday's observations led to the discovery that electric and magnetic fields are connected. Faraday interpreted the relation by stating a law: *"Electric fields are generated by changing the magnetic fields with the time."* We amend it with adding the space in the law as: *"Electric fields are generated by changing the electric or magnetic fields with the space and time."* By making this amendment, we clarify that these three cases originate a 12 point connected SST connections that introduces EMF in a (second) circuit for electrons to flow and have a current in it.

It is the electric field that drives the electron around the wire—producing the EMF in a stationary circuit to have an induction current—when there is a changing magnetic flux. Maxwell represented these facts in a set of three partial differential equations. These equations are known as Faraday's law and mathematically expressed as

$$C \, \nabla \times E = - \frac{\partial H}{\partial t} \qquad (3.7.1)$$

These innocent looking partial differential Equations (3.7.1) hide three mathematical facts related to the motions of electrons. First: an electric field exists in a wire when there are electromagnetic changes in the space and time—as noted by Maxwell as changes in the neighborhood. If one of the pairs of wires has change in current, a current is induced in the other, or if a magnet is moved in and out of an electric circuit, there is a current in the circuit. These changes are not necessarily limited to be a uniform (change) in their motion. These changes induce currents in the circuit that are tangible due to the real EMF produced by the non-uniform motions, and non-fictitious forces that appear with the non-uniform motions of the reference frames in Newtonian mechanics. This is the fundamental difference in phenomena of Maxwell's electrodynamics which differ from the Newtonian dynamics. Based on the noted differences, one cannot extend the mathematical representation, like the GPR, from Newtonian Mechanics to Maxwell's electrodynamics, as long as the space and time are absolute and independent.

There was a gap, a limitation in the mathematical representations of these two phenomena. It is the union of the space and time, which was there in the Equation (3.7.1) and revealed here in the above presentation, that also appeared first time in the STR, provides the transition of GPR from the Newtonian mechanics to Maxwell's equations, and the gap was removed.

Second: it appears that the Equations (3.7.1) do not follow the GPR at a first site until we recognize that the $C$ associated with the phenomena needs to be considered a constant and the space and time to be united. Under these conditions it follows from that the complete set of Maxwell's equations satisfies the GPR. One can see its details in any book on STR.

Third: the Equations (3.7.1) establish a relation between the changes in the electric field in the space with the changes in the magnetic field with the time. These changes show that the electric and magnetic fields, and the space and time are not absolute, but they are interconnected and permit to have changes in one with respect to the space to associate the modification in the other with the time. All these changes remain intact under the constancy of velocity of light and the union of space and time suggest that the matter connected with these fields is also associated with the time.

The Faraday's experiments identified the invisible pattern of the physical world and derived it as Faraday's law of induction, which Maxwell expressed in the mathematical Equations (3.7.1).

The set of eight Maxwell's equations are simple (presented in Chapter V), which describe the complete motions of eight points connected electrons with all possible changes in electric and magnetic fields and associated currents in the space and time, and include the velocity of light in their presentation. Due to the presence of the constancy of velocity of light in the Maxwell's (wave) equations, it is thought that it includes the light waves, and describes the motion of photons. If this is the case, the Maxwell's wave equations should explain the blackbody radiation of quantum mechanics and other properties related to particle nature of photons; but it does not. So, from these observations it follows that the Maxwell's wave equations do not represent all the physical properties of photons—light, except that they incorporate the velocity of light; the ratio of electrostatic electrodynamic units—which happens to be equal to the velocity of light.

In the observational identifications of Faraday's and Maxwell's derivation of the equations, motions of the circuits or the magnets, or the changes in the currents are not restricted to comply with the uniform motions. There is no restriction on the motion of the circuits, magnets, or the changes in the current, which is one of the fundamental requirements for the GPR and the STR. This is new and shows here the fundamental difference associated with the electromagnetic fields. By accepting the condition of the union of space-time, one can demonstrate that Faraday's law of induced current meets the GPR. Once we accept the union of the space-time, the Maxwell's equations under the non-uniform motions meet the GPR.

As a fact, following the STR calculation, one can verify that the velocity $V$ (which is equal to that of light) does not change its value by change in the velocity $v$.

The set of eight Maxwell's equations describe the complete motions of eight points connected electrons with all possible changes in electric and magnetic fields and associated currents in the space and time. The set of eight equations include the partial differential Equations (3.7.1) that describe partial changes of electric and magnetic fields and changes in the current and induction currents.

The representation of Faraday's experiments in Equations (3.7.1) depends on twelve points of SST connections of the electrons, which are mathematically not considered and not represented in the Maxwell's equations. These twelve points connected electrons in their motion are constrained to remain within the atomic shell, but originate an induced current in a circuit at rest. The Equations (3.7.1) establish the relation between the changes in the fields and the associated currents without revealing the added four points of spatial-connection in Maxwell's equations.

The added four points of spatial-connection remains hidden in the Maxwell's electrodynamics as it considers the motion independent of the added requirements—constraints—that the electrons' motions to remain within the atomic shell, and treats the space and time absolute like the one considered in Newtonian mechanics. In the space-time manifold of the electrons' motion in an atomic shell, the noted constrains and added points of connection, each has a four points of spatial-connection; thus, these added spatial-connection and constraints remain latent without violating the Maxwell's equations.

## 3.8 Explanation: Why velocity of light maintains constancy under a relative uniform velocity of motion

An observation related to a uniform velocity of a moving object depends upon how the velocity is reported by an observer moving with a uniform velocity. The resultant velocity observed by an observer is equal to a resultant of two velocities, which is equal to the calculated velocity derived through the laws of addition of two (Newtonian velocity) vectors. Let us consider a simple example of velocity of a flying plane. When a plane flies with uniform velocity **X** in sky with wind velocity **Y**, where both the velocities are parallel, and then the plane's resultant velocity is different depending upon how the plane is flying. We observe that the plane's head-wind velocity is equal to (**X-Y**), and tail-wind velocity is equal to (**X+Y**). The resultant velocity depends upon how the plane is flying. In this case, all of the different velocities appear in the absolute (Newtonian) space and time, and have 2 points of connections.

But, for the velocity of light "**C**" this is not the case. A change in velocity of light disappears and, it is constant. This constancy has created a controversy, as SRT did not clarify its true physical connections. The discussion and representations of Sections 3.6 and 3.7, partially clarifies our understanding that the velocity **C** is constant due to its higher number of points of connection with the space and time.

For the controversy over the constancy of velocity of light in physics when discussed between two observers or along emitting and receiving objects that are moving with uniform relative velocity, a few facts we are ignoring: the roles of the space and time appearing with the Maxwell's electrodynamics are different from that of the Newtonian mechanics, and there is a need to recognize and summarize their contribution in motions. The space and time accommodate the motions of matter of electrons and gravitational bodies, but in each case the associated matter and its velocity act differently, and have different number of points of connections in motion.

For the constancy of velocity of light C appearing with the Maxwell's equations, we note here that, has gained a status of Triple-Crown property in physics: It is a constant, it unites the space and time, and it does not alter when observed under relative uniform motions.

The first two jewels are, respectively, presented in Sections 3.5 and 3.4. First, the constancy of velocity of light is required to maintain the Gaussian units of transformation and the continuity of one of the Maxwell's equations. Second, the union is already established by the electromagnetic field associated with the motion of electrons, magnets and associated electromagnetic properties.

Third, the velocity of light does not change when observed under the uniform relative velocity between two observers or between the emitting and receiving entities. This fact can easily be derived and verified by simple mathematical calculations as presented in many books of Special Theory of Relativity (STR). The STR calculation was derived to satisfy the empirically observed reality of Michelson and Morley experiments and maintaining the Galilean principle of relativity. The mathematical calculations and empirical observations, however, have physical reasoning why this is the case for the invariance of velocity of light. This case is similar to the case of moving magnets, before Faraday's experimental presentation, as reported in Section 3.7 that does not produce an electrical current in a straight conductor. In both the cases, in the space and time connected manifold, an electric current and velocity of light have higher number of points of connections. The fewer point connected magnet and the Newtonian velocity cannot influence the higher points connected electrical current and the constant velocity of light. Thus, the (two points connected) relative velocity $v$ (with absolute space and time) cannot change the $C$ (the eight points space-time connected), the constant velocity of light.

## 3.9    Conclusion

1. Time arises from a continuous motion of matter in space. The changes in the direction of time are from the past to the present and from present to the future. Only the present is observable. The time does not have its own origin as is the case for the space that can be expressed with a selected coordinate system in the selected reference frame. In motions of electrons, their electromagnetic flux velocity (or the velocity of light) $C$ unites the space and time. For a continuous space and time, it is necessary and sufficient that the velocity $C$ must be constant.
2. In electrodynamics, Ampere's law of magnetic induction upholds the union of the space and time through the constant velocity of light.
3. The velocity of light has eight points of spatial-connection.
4. From Faraday's experimental observations it follows that there is an induced current in a circuit when there is a relative motion in a second circuit carrying a steady state current, or there is

change in the current in the second circuit when a magnetic pole is introduced in or out of a circuit.

5. Faraday's experiments are examples of non-uniform relative motions between two circuits, or a circuit and magnet to generate real forces (EMF) to move the electrons producing a flow of current, contrary to the Newtonian mechanics having fictitious forces for the non-inertial frames.

6. Faraday's law of induced current meets the GPR under the union of space-time as shown in the STR. And under the same union the GPR reduces to a special case of the Faraday's law.

7. The velocity of light $C$ is constant and it does not change with observers or its emitting and receiving objects are moving with uniform relative velocity $v$. A fundamental reason to defy the Newtonian concept of addition of two velocities is that the velocity of light $C$ unites the space and time and has eight points of connection, while the Newtonian relative $v$ velocity has only two (spatial) points of connections, and the space and time are absolute and independent of each other.

# Chapter IV

## Electrons Reign in Space, Time, and Matter

### How?

The strength of royalty emanates (arises) from experiments.

We need to step out of the old ruts for new insights to rein the chaos.

### Why?

Continuity and discontinuity transcend in our finite understanding; the latter demands an individuality and freedom; the former establishes connection between the individuals and unions; the process of dealing with them provides progress in science.

## 4.1    Introduction

It can virtually happen to any scientist while performing an experiment or establishing a theory or proposing a conjecture based on known theory on a specific motion of matter or its characteristics, but other scientists learn later that a specific material behavior, like that of an electron, turns out to be opposite to a forecasted prediction. Most of the time, this is a matter of degree of embellishment, something that can be countered by skepticism of freshly observed facts, forcing to make the changes. This is the case for charged particle—electron. As reported in Appendix A, a simple observed fact of spinning of a paper clip in front of the magnetic field—a cause and effect from the motions of electrons in the associated materials—leads us to explain the half spin of electrons in this chapter. In order to keep up with old terminology for electrons we will use the words "electrical charges" or "charged particles".

In Coulomb's law electrons appeared as true evangelical (Newtonian) particles. But when thoroughly analyzed and further studied, the charged particles turned out not to be a choir boy of Newtonian mechanics, though not visible, but a tough individual to standup at the foundation of the law with loaded individuality demanding to have two types of identity—positive and negative charged particles. These two independent set of charges maintained equilibrium—namely, the generation of a positive charge is accompanied by the generation of an equal negative charge as the law of conservation of charges, similar to the Laws of conservation (presented in Section 2.0 below) of Newtonian mechanics.

Newton and other scientists noted an apparent connection between electrical charges and the hostile environment of lighting in the sky. But they did not know how these particles can be studied as they were invisible.

It is the freedom of the charges that demands the understanding of their liberty or that can easily give mankind a death if not followed with respect! One of the American Freedom fighters—Benjamin Franklin (1706-1790)—noted this fact with respect, and experimentally proved it for these free electrons by flying a kite to the sky, while taking a risk on the life of his own son—William Franklin (circa 1730-1813)!

Franklin observed that the point charged free electrons can easily be realized in transferring from a small electrified iron ball to the pointed piece of metal when brought near the ball, but not by a blunt piece of metal. Due to this reason, it is possible to draw the electrical charges from electrified cloud on pointed metal ends attached to a long metal rod running from the top near the cloud to the ground. Thus, considering the free charged particles, experimentally, Franklin snatched the lighting from the sky by his metal rod scepter and kicks the death demeanor to the ground.

How could the concept of particle charges have continued to happen? The particle physicists, including Newton, should have not professed any further the particle characteristics of charges. But, the physical theological argument of particle has still continued, as no easy alternate was available and, even today, it has appeared in some of the physical theories as particle—point charge. In this chapter we will work to put it to the rest.

Electrons are enigmatic! An electron likes to act in exception to the rule. An electron is not normal, not a member of typically studied material of Newtonian mechanics, not easy to comprehend its characteristics, and many times it behaves contrarily to the conventional wisdom. We have always found corrective encounters whenever we have attempted to declaim too haughtily on what is or what is not normal in the theoretical and practical studies about electrons. In the following sections we will see what was known at the time of Coulomb's experiment from the observations and theoretical understandings the motion of planets, ideas of action at a distance and certainty in the Newtonian mechanics, how do they differ in the Maxwell's electrodynamics, and how does spin of electron differ from the conventional spin of earth and other planets.

The e of electrons appears in space, time, and matter (STM). An electron acts according to its connection with the STM. Before we start discussing on the electrons, we summarize activities of 'e' in the following poem:

## e stands for electron

I am in the beginning of an end
And I am in the end of space and time

I surround every place and established time
And every place and established time are in me

I am in matter, and the matter is in me
And e-motion connects the space, time and matter

e of space, time and matter fall between the edge
And edge of all spectrums of transformations falls in me

I unite the space, time and matter
And the space, time and matter unite with me

I am essential to excitement of changes
I am the electron and 'e' stands for me.

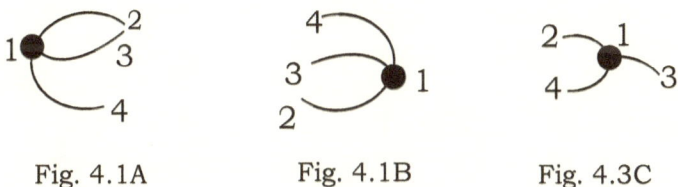

Fig. 4.1A          Fig. 4.1B          Fig. 4.3C

Figure 4.1A represents (partially) the almost closed loop of e for electron, but it is really not closed. It corresponds to a set of three perpendicular curvilinear curves—can be considered as 3 axes with origin at a material point 1 that satisfies the four-point connection of the electron at rest. The first one from possible figures easily takes us to consider it to be an e. One can make other geometrical presentations of e as shown in Figures 4.1B, 4.1C, and so on. Similarly, there exist other geometrical representations and formulations for the electrons, out of which we will select a suitable one meeting the requirements of the discussion on hand, or the model which explains the observations of phenomena being studied.

## 4.2    Review of Post-Newtonian era developments

Newtonian laws of motion to study planetary motions and a set of simple equations to solve the sophisticated complex problems dominated the three centuries of the post-Newtonian era. Many mathematicians and theoretical physicists like Joseph-Lois Comte de Lagrange (1736–1813), Leonhard Euler (1707–1783), Pierre Simon Laplace (1749–1827), Jean le Rond D'Alembert (1717–1783), Simeon Denis Poisson (1781–1840), Carl Gustav Jacob Jacobi (1804–1851), William Rowan Hamilton (1805–1865), and others were associated with developments of mechanics. And their contributions are far interesting and too much to report about, but we will present a brief summary of their results useful for the ideas we want to develop here for electrons.

Two sets of general principles—conservation principles and minimal principles—steered these men in their contributions to post-Newtonian mechanics. They developed three conservation principles, namely, the principle of conservation of momentum, the principle of conservation of energy, and principle of conservation of angular momentum. These principles, associated with the social material things and religious theology requirements, provided to study what remain constant and what items do not get lost during the changes in the motions. The minimal principles on the other hand impose conditions on some measurable quantities—let's say action (which we will define in the following)—or the energy associated with events of motion takes place in such a way that the trajectory of particle will occur along that path in which these quantities are, as smallest as possible, minimal. These principles simplified to study the motions of particles, bodies, system of particles, etc. analytically without going into their force configuration details.

## Concept of certainty

The success of Newtonian mechanics in applications of the effect of gravity led Laplace to believe in the reasoning of a completely deterministic universe. Based on this belief and successful applications to predictions of planetary motions, Laplace and others thought that the known scientific laws would allow humans to predict everything about what would happen in the future about a dynamical system, if we completely knew the state of the system—its position and velocity—at the beginning of the motion. For example, by knowing the location of the moon and its velocity along its orbit at a given time, one can accurately predict rise and fall of the tides on earth at a future time, with the repeatedly correct forecasts. Predictions with certainty looked obvious for similar cases for the motion of planets, projectiles, and other similar mechanical systems. Many people resisted this doctrine of certainty at that time based on religious grounds. On the other hand, to follow the certainty in predicting the motion of particles, particularly that of electrons and photons, became prominently important among physicists of the early-20th century during the development, understanding and acceptance of uncertainty principle of quantum mechanics. The uncertainty principle rose from an extension of Hamilton's least action principle which was developed for the particles that followed the laws of Newtonian mechanics. Let us first review the contributions of Lagrange, in particular for the minimal principle. From this work, Hamilton modified the Lagrangian framework to introduce Hamilton's least action principle. Their work is summarized in the followings.

## Lagrange's Contributions to the study of motion of system of particles

Newton's equations of motion explicitly deal with the motions of individual particles subject to the interaction from other particles and the external forces. There are three equations of motion for each particle. And when there are three or more particles in motion it becomes difficult and complicated to solve the equations to describe the motion of these particles under their reactions and external forces. The problem becomes hard and impossible to handle to solve a large number of particles. To get rid of the noted difficulty associated with the Newtonian mechanics, Lagrange introduced a procedure to solve the problem for the system of particles in motion by introducing the concept of their degrees of freedom and association in reducing the number of equations of motions by considering them as ensemble of particles.

The method of Lagrange provides us a systematic procedure to develop and derive the equations of motion of a dynamical system, without going through the rigorous representation of how the forces, action, and reactions are acting in the same dynamical system. Loosely speaking, these methods direct us to derive the kinetic and potential energies of the system in terms of the degrees of freedom, and then perform certain mathematical operations on these two functions, and derive the equations of motion. Once the differential equations of motions are derived, one can find their trajectory from integrating these equations. This method is very helpful to carry out the primary task of dynamics—namely, solving dynamical problems—to find out how the systems move.

To summarize the Lagrange method, consider the motion of a single particle. It is free to move in any one of three perpendicular directions, if there is no force acting on it, or will move in a direction of force that can be expressed with three equations represented under a reference frame of three perpendicular axes with three coordinates. We can say that the particle has three degrees of freedom. Three equations are required for each such particle in an ensemble, and the total number of equations becomes forbiddingly large as the number of particles increases, as was the case of Newtonian mechanics. To reduce the number of old coordinates associated with the particles, Lagrange introduced the known algebraic and geometrical relation on the old coordinates of the ensemble of particles. By introducing the new coordinates he reduced number of coordinates and named them "Generalized Coordinates (GC)." The particles have GC and generalized velocities. Then he introduced the function of the system, known as Lagrangian, consisting of its kinetic energy minus its potential energy and expressed in terms of the new generalized coordinates, and its velocities. The equations of motion are derived from the Lagrangian. Based on the nature of Lagrangian, there are classifications of all dynamical systems as—scleronomic or rheonomic; conservative or non-conservative; holonomic, or non-holonomic. For the dynamic systems classification, the reader is requested to refer a book on dynamics proper.

## Hamilton's Contributions on the least action principle

The introductions of laws of conservation principles, followed from Newton's laws of motion, are equivalent to provide deeper meaning of the laws of motion. The conserved quantities are algebraically derived from the elementary physical entities. And these quantities fall into two classes. Some quantities are vectors—like momentum, angular momentum—that depend on the reference frames; while the others are scalars—like energy, action—that do not depend upon the reference frames, and have desirable properties to study further.

In the seventeenth century, Pierre de Fermat (circa 1601-1665) introduced the Minimal Principles in physics to study the propagation of light, based on the "principle of economy," a scalar quantity, according to which an event in nature unfolds in a shortest possible time, and explained the reflection and refraction of light. In the 18th century, the French mathematician Pierre Louis Maupertuis (1698–1759) applied the idea of the principle of economy of nature to Newtonian mechanics, by extending it through the concept of action. Maupertuis' definition of the action was flawed and Hamilton corrected it and introduced the "Principle of least action."

The principle of least action had played a major role in the developments in the post-Newtonian era. If a physicist or a mathematician or an engineer needs to overcome his nostalgia for the practical and does not want to embark on the study of Hamiltonian equations derived on the basis of the principle of least action having real abstraction of the motion of particles, he may be surprised and rewarded by having at disposal of a powerful tool to study of electricity, magnetism, optics, quantum mechanics, etc. which do not represent dynamical systems in the ordinary mechanical sense, but have behaved that way and provided a major contribution to these disciplines. The foundation of the least action principle goes back to the Newtonian mechanics.

In Newtonian mechanics there are two measurable things attached with a particle; its position and its momentum in the motion. If these two entities are known at any point of the trajectory, the complete orbital details of the particle motion can be deduced from Newton's second laws of motion through the knowledge of the forces acting on the particle. Since, over an infinitesimally small stretch of the particle's path, its momentum is practically constant. This constant momentum multiplied by its infinitesimal path of the orbit is called the action. The action is a scalar quantity, and changes with its position along the trajectory. According to Hamilton, in all possible motions of a particle from a starting point to the end point, the particle chooses that path along which the action is least, minimum. This is the "principle of least action." This principle abolished the action at a distance embedded in Newton's law of gravity, which was abhorred (disliked) by many including Newton.

As noted before, the Lagrangian of a particle is its kinetic energy minus it potential energy, and can be expressed as a function of generalized coordinates and its velocities and time. Hamilton used the concept of action as force field in which the particles are moving under their potential energies. Hamilton derived the generalized momentum from the Lagrangian, and constructed a function, called "Hamiltonian function or simply Hamiltonian." For a dynamical system, the Hamiltonian is equal to the sum of kinetic and potential energy of the system. The Hamiltonian converts the n second degree differential equations of the system into 2n first degree differential equation for the trajectory of the system. There are many benefits to follow the Hamilton's method to study the motion of particles. But, there is no space here in this book to enter into its applications of general methods of dynamics. The above remarks are intended to summarize what had happened in the developments of physics and mathematics, and has provided major and very valuable contributions in the development of physical theories.

## Comments on Lagrange's and Hamilton's contributions

Lagrange's and Hamilton's contributions moved the study of dynamics by analyzing a system analytically, and are far superior to the old methods, and simplified the process of understanding the motion of system of particles. But these approaches also have limitations. We will note only those limitations of the concept of least action that are applicable to electrons in the following:

First, it is assumed that the phenomena of motion is continuous in space and time, and particle's location and its momentum are well defined, and can be measured simultaneously at a point on its orbit. This is not always the case for non-Newtonian particles. These particles may be moving in non-continuous space or the space and time are connected as space-time manifold, or the simultaneous measurements of the particle location and momentum are not feasible, as these particles in motion are not readily visible and available simultaneously to locate their positions and their momentums.

Second, the action is based on the location of a particle, and the particle's geometrical and material locations are the same. In Chapter I we noted that these two points are not necessarily the same, and considered as two separate entities. We will see that this is the case for electron, where it has two points and simultaneously measuring its position and momentum is not always feasible.

Third, the measurement of particle location and its momentum presupposes that these entities are infinitely divisible, so that these quantities are always observable and measurable. This is not feasible if the matter is non-divisible, so that either one of them is possible to measure or not the both as is the case of photons, which have finite size based on the quantum mechanics, which cannot be divisible less than a Planck constant $h$, leading to the uncertainty principle.

Fourth, one cannot define the generalized coordinates and their velocities for the Lagrangian for the electrons moving in electromagnetic and magnetic fields, as was the case for Maxwell in derivation of the Maxwell's equations. Without the Lagrangian, we cannot have the true canonical equations of motion for the Hamiltonian to apply the least action principle.

Fifth, the action is a product of two vectors—displacement of position of particle and its momentum. As we noted before, an electron has four points of connection, and when it moves, it adds extra points of connection, making the momentum vector (associated with the field) to be on the opposite side of the surface where the (material of) electron is located. The election's position and momentum, both cannot be measured simultaneously, and so one cannot define the action and so cannot apply the principle of least action to the motion of electrons. We will see in Section 4.4 that the magnetic field originated by the motion of electron is moving back and forth and remain on the opposite sides of the surface on which the (material of) electron is located.

### Developments of other branches of Physics

The other branches of physics, such as electricity, magnetism, fluid mechanics, thermodynamics, optics, and others, grew slowly, but developed independently of Newtonian mechanics. These branches of physics required to do experiments in laboratories with very little resources to verify the findings to understand the technical and complex nature of matter. Most of the experimental works were done on trial and error basis. Experiments were performed with crude equipment and had no mathematical basis (from which) to record and develop the data or a consistent theory.

Sometime the results appeared accidently, but overall the progress was sluggish and with doubts. But as the saying goes: "Slow and steady wins the race." The experimentally observed properties of matter associated with the charged particles in motion provided better understanding of various possible connections of electrons, and that of the space and time; some of them are presented in Chapters II and III earlier.

### Oersted's observations of magnetic fields in space due to currents

During an experimental demonstration, Oersted observed that a magnetic compass needle originally set parallel to a wire along the north-south direction turns by 90 degree facing east-west direction when the wire connected to the positive and negative ends of a voltaic cell and current remained flowing through the wire. He was puzzled and surprised to observe the magnetic effect due to the current in the wire. He changed the direction of flow and observed that the magnetic needle turned in opposite direction. He also observed and reported that the magnetic field was in a circle around the wire. The magnetic effect in the space due to the current was one of the greatest scientific discoveries of all time. The production of magnetic field, due to motion of charges, acts and rotates the magnetic needle differently than the one studied under the influence from electrostatic, magnetostatic, and gravitational forces.

A similar observation for the rotating paper clip due to magnetic field, reported in the Appendix A, forced us to look into: Why we have the spinning of the earth? We answered the question in Chapter I by introducing an improved definition of a particle. How can one demonstrate that the electrons spin without using the tools of quantum mechanics? How to demonstrate that the magnetic field is couple (with a torque) and not a force? One can observe answers to these questions by performing a simple experiment presented in Appendix A.

The rotation of the magnetic needle showed that it is caused by a torque—having two equal and opposite forces at a distance—originated by the electrical current (from motion of electrons) and not a simple force of attraction or repulsion, as it appeared in the production of a force originated by two masses, two electrical charges, or two magnets acting on each other. Plus, note that the force of attraction or repulsion is at a point, along a line and not separated to originate a torque, while in the Oersted's experiment, he observed a torque.

The rotation of a magnetic needle shows that electrical current produces a magnetic field, and the magnetic field is at a right angle to the electrical field. These two fields are at right angle as long as he electrical charges—electrons—are moving. From this observation, it was inferred that the magnets and magnetic fields are due to the motion of electrical charges.

## Comments on Oersted's Contributions

During the era of Oersted, electricity and magnetism were considered as two separate entities. It was thought that magnetism was associated with magnet or with magnetic pole, and not with the electrical charges. The observation of production of magnetism by the electrical current was hard to explain. The experiment appeared as a paradox. But the introduction of a concept of the field by Faraday made it easy to explain these observations and later permitted Maxwell to unify both of them.

We will follow the concept of the fields, with understanding that it does not permit to observe all possible motions of electrical charges and will not allow realizing its limitation, as was the case in using a concept of gravitational field in understanding the orbital motion of earth that did not permit to understand the observed spinning of earth. The concept of the field does not portray a complete picture of three items associated with the motion of electrical charges—the electric field, magnetic field, and the motion of electrical charges. The electrical charges move along the wire and the magnetism produced by the current is in a plane surface enclosed by a circle. The circular surface can be represented by three points passing through the wire and is perpendicular to the wire. Thus, Oersted's observation represents four points connected entity, which appeared for the first time after Coulomb's experiment. This was one of the reasons why Oersted's observation appeared contradictory. Oersted's experiment reveals four points of connection, in a sense it is limited from the available six points of connection of the electromagnetic induction that Faraday extended and eight points of connection available for the electric and magnetic fields that Maxwell extended.

## Faraday's contributions on electromagnetic Induction

Michael Faraday had no formal education to work in the scientific community, but had abilities to self-learning, physical intuition, minute observations, attention to details, love of humanity, fond of children, and characteristics to tolerate others' criticism. He was one of the noble scientists in the development of physics and chemistry to accelerate by simultaneously working with well thought-out experiments and, as he had scarcely any training in mathematics so without using equations but had insight into nature to compensate the deficiency, theoretically presented his findings in methodically thorough details that were easy to understand.

Faraday started his carrier as a bottle washing assistant in a chemistry lab, and in his spare time attended to the lectures delivered at the Royal Institution, London. In the lectures, he heard about the details of Oersted's experiments showing that an electric current deflects a magnetic needle. These facts were important to Faraday, but as usual, he cross-examined the assertion and verified the facts as his own. Once understood the facts of Oersted's experiment, Faraday looked in the opposite direction to understand the effect of magnetic force on the electric current. Faraday was aware of his deficiency in mathematics that forced him to represent the physical phenomena in terms of physical model and developed the model of "tubes of forces."

The "tubes of forces" model led Faraday to the discovery and explanation of the electromagnetic induction as a cutting of the tubes of forces. A conductor wire coil has free electrons to move. Whenever tubes of forces are cut by a conductor coil, a current flows in the coil; the faster the tubes are cut, or faster the tubes are moved, the greater is the electromotive force (EMF) that is induced in the conductor to move the electrical charges (electrons) to produce current in the coil. The details of these findings are discussed in Chapter III.

## Comments on Faraday's contributions

According to Newtonian mechanics, the force is along a line and starts from a point, and it is due to an action from other particle at a distance. If anyone should support this concept of force, he should be the one who has good mathematical background, or has studied physics in depth, or using it daily in (mechanical) engineering. But that was not the case with Faraday and Maxwell. Maxwell, who had passed tripos and awarded the second position as a wrangler—the highest and most difficult to achieve the mathematical honors at Cambridge—and has made large contribution in physics, did not use this concept of force. By supporting Faraday, Maxwell developed his electrodynamics and derived the Maxwell's equations.

It is time for us to acknowledge the sad truth about the forces appearing in Newtonian mechanics that needed to be overhauled. And Faraday did it with the concept of "tubes of forces" that replaced the concept of force at a distance, and enlarged the concept of field, and Maxwell generalized it.

The tubes of forces created a system of forces in the neighborhood, encircled by a surface, consisting of three points of connection. Similarly, the wire loop coil is also a system of three points connected surface with free electrical charges (electrons). And the motions of intersection with either the tubes of forces or the wire loop coil originates a six point connected domain for the EMF to have the electromagnetic induction and the current in the coil.

Faraday showed that electricity and magnetism are reciprocal phenomena, and under the suitable conditions, each one of them can induce the other, and discovered a relationship between them (and light).

**Maxwell's contributions on Electrodynamics**

Faraday's experimental discovery of electromagnetic induction phenomena, the development of concept of tubes of forces, with lack of mathematical knowhow, the task of unification of electric and magnetic fields were left to Maxwell.

Maxwell developed a dynamical theory of electromagnetic field to unify the electric and magnetic fields in a set of eight equations (presented in Chapter V) that accommodate the changes of the noted fields from point to point in their neighborhood. The equations establish a relation of the changes in one of the fields at a point with the other field at the same point in space and time. Maxwell did this unification by introducing the concept of interaction of the charged bodies depends upon the surrounding space, and forces between two bodies depend not only on their neighboring positions, but also on their relative velocities.

It was not customary during Maxwell's era, though developed by Lagrange and Hamilton, to think in terms of abstract ways to solve the dynamical problems using the concepts of Lagrangian, Hamiltonian, and fields. So, Maxwell developed his electromagnetic theory in terms of a mechanical model—oscillations of a pendulum—a condenser in which space was considered filled with elastic solid, namely ether (that is not required now anymore, as is the case after Einstein introduced the union of space-time through the constancy of velocity of light in the Special Theory of Relativity). During the Maxwell's era, there was hesitation and sometime resistance to accept Maxwell's mechanical analogical model, and there was no experimental justification available to demonstrate the theory. Today, we have plenty of examples to substantiate his theory and have many experiments to confirm the Maxwell's equations.

To synthesize the electric and magnetic fields, and to satisfy the divergence of the magnetic field produced by the electric currents, we will summarize the presentation of Maxwell. Let us consider electric field in a condenser of two metal plates at a small distance apart, with one of the plates grounded and the other charged with negative charges on the plate, producing electric field lines perpendicular to the plates with positive charge on the other plate. Each of the plates is a three point connected surface and the electric field between the space has a total of six points of connection. The electric field has a potential to produce a current if joined by a conductor. According to Maxwell's model, the electric field in the condenser denotes the largest potential energy, similar to the one in a mechanical oscillator, like the one in a pendulum at a highest point of its oscillating motion.

Let us analyze the changes taking place in electrical and magnetic fields, by paying particular attention to the events taking place with an experiment performed with the condenser described above, and summarize the observations as follows:

First, connect the two plates by a conductor (metal wire) permitting electrical current to flow in it. The negative charges from one of the plates flow through the conductor, originating a current outside of the condenser. As the current flows, the electric field between the space of plates changes and charges move from one plate to the other till it makes both the plates charge-less, zero charges, and no current flows through the conductor.

Second, Maxwell named the current passing through the conductor as "displacement current". The displacement current will originate magnetic field in the neighborhood of the conductor. The displacement current permits to have zero value for the divergence of the magnetic field originated by any steady state flow of electric current and also due to the changing currents in the neighborhood. In this case, the flux of current out of any closed surface is equal to the decrease of the charges inside the surface, which is interpreted as the conservation of charges; this fact is represented mathematically as the last of the Maxwell's equations.

Third, the current flowing in the conductor produces circular magnetic lines of forces surrounding the conductor, as reported by Oersted, that changes from zero to maximum, while the charges deplete reducing the electric field from maximum to zero in the condenser and in the conductor. The magnetic field is produced in the space outside the condenser due to motion of the charges, and denotes the kinetic energy related to the charges, similar to the one in a mechanical oscillator, like the one in pendulum at the lowest point in its oscillating motion with highest kinetic energy.

Fourth, the neighboring space of the condenser has a maximum magnetic field, and it starts going down as there is no current to sustain it. The changes in the magnetic field from maximum to reduce it to zero produces electric field in the conductor, as described by Faraday of producing the EMF (electromagnetic forces), and to the condenser. The produced current will be flowing in the opposite direction of the initial current case described in the first observation, and will have charges again in the condenser, having the potential energy noted before.

Fifth, the whole process repeats itself; the phenomena of changing electric and magnetic fields goes on periodically over and over again, like the mechanical oscillations with a periodic time and frequency of a pendulum. The model experiment shows analogy of the mechanical and electromagnetic oscillator.

Sixth, the electric oscillator of the condenser does not continue forever, but ceases after a certain time as is the case for a pendulum. In the process, energy is lost in both the cases. For the pendulum, the energy loss appears, due to friction, as heat and for the electromagnetic oscillator the energy loss appears, very quick, as radiation.

Maxwell's electromagnetic theoretical model provided a simple physical description from the oscillating condenser, looked superficial, and physicists and mathematicians of the era were skeptic and not ready to accept it. To compound the problem, the developed theoretical model predicted the existence of electromagnetic waves travelling with the velocity of light, which can be rigorously derived from the six equations of Maxwell's electromagnetic field theory. A set of three equations for the two—electrical and magnetic—fields, representing three components of the field in three spatial directions of the field, shows how they, respectively, change with the variations of the changes of the magnetic and electric fields from point to point. These two sets of equations show that the electric and magnetic fields are propagated together as periodic oscillations at right angle to each other and are at right angle to the direction of the propagation of the (transversal) waves. The propagation of the electric and magnetic field waves are 90 degrees out of phase, as we noted in the second and third observations above. Thus, the waves have maximum electric field when the magnetic field is zero, and vice versa.

The reservations in the correctness of the model unifying the electric and magnetic fields developed by Maxwell's theory persisted for some time, as there were no instruments available, and experiment conceivable to verify the noted oscillation predictions. All the reservations related to theory's correctness were removed in December 1888 by Hertz through the production of electromagnetic waves in electrical oscillator apparatus and their reception in the same kind of oscillators located at various places in the space, away from the source of the waves' production.

Hertz produced the electromagnetic waves by discharging a condenser consisting of, instead of two metal plates, two metal spheres separated at small distance with high potential difference between them, creating discharge in the form of spark which originated rapid electromagnetic oscillations. The waves thus produced created spark to jump between two similar spheres, and two wire meshes connected by a similar conductor placed at a distance. He showed the electromagnetic waves' transverse nature through experiments on reflection, refraction, diffraction and polarization. In short, he completely established the electromagnetic wave characteristics as predicted by the Maxwell's equations and opened the gate for the radio technology.

## Comments on Maxwell's equations and Electromagnetic waves

1.  The set of condenser and the conductor lacks the symmetry between the electric and magnetic fields appearing in them, and have abrupt discontinuity in its model of oscillations; neither of them directly participate with the either of the fields or with the their changes in the space. On the other hand, an introduction of "displacement current" permits to satisfy the divergence of the magnetic field produced by the motion of electrical charges, and currents in the space; and permits to meet the conservation of the charges and currents in the space. This was Maxwell's ingenious idea in establishing the union of electrical and magnetic fields.

2.  The concept of displacement current does not participate in the derivation of the electromagnetic field and also in its wave nature.

3.  The electromagnetic waves are derived with only six equations out of eight Maxwell's equations as noted in Chapters II and III.

4.  The electromagnetic waves have six points of connections and not the eight available for motions of electrical charges (electrons).

5.  The developed electromagnetic waves do not include all possible motions of electrons in the space, which we will report in the following.

6.  It is a coincidence that all of the electromagnetic waves are travelling with the velocity of light, which is almost equal to $3.00 \times 10^8$ m/s in vacuum. The electromagnetic waves exist with enormous range of frequencies, starting less than $10^4$ Hz to greater than $10^{22}$ Hz, known as electromagnetic spectrum. Historically, the spectrum of electromagnetic waves have been called radio waves (includes A.M., F.M., and microwaves), infrared, visible light waves, ultraviolet waves, X-ray waves,

Gamma rays and so on. The boundaries between different classes of waves are loose and overlap in practice.

7.  Out of the large range of electromagnetic wave spectrum, visible light falls between a small range of 4.0 x $10^{14}$ Hz and 7.9 x $10^{14}$ Hz. (For details refer to any book in physics, like *Physics, 2nd edition* by Cut-Nell and Johnson, pp. 690–691.)

8.  Though light is considered a part of the electromagnetic radiation, its geometrical and physical structures are different from that of other electromagnetic waves.

9.  The waves of electromagnetic radiation have mostly six points of connections, while the waves of light and photons have higher number of points of connections.

We will not discuss the properties and connections of photons in this book as its motion and quantum mechanics fall outside the scope of the book. But, in the following we will discuss the motions of electrons satisfying the Maxwell's equation taking place in its neighborhood represented by the electromagnetic waves, and reveal why the observed changes of oscillations take place in the electric and magnetic fields.

## 4.3 Spins of particles, moon, earth and other objects moving in a plane

Both Kepler's and Newton's laws are based on the motion of planets—particles—in the space. Kepler's three laws state: (1) Planets paths are ellipse; (2) The line joining a planet to the sun sweeps an equal area in equal time; (3) For a planet, the periodic time to complete the orbital motion around the sun is proportional to the three and halves power of the orbital radius (average distance of the planet from the sun).

Newton's three laws of motion are: (1) Particle moves under a uniform velocity (including at rest) unless an external force acts on it; (2) Particle acceleration is proportional to the force acting on it; (3) Action and reaction are equal and opposite and act along a straight line.

These laws are for the particles and are limited to simple two dimensional plane motions—permitting the maximum of three points of connection, as noted in Chapter I. All geometrical properties of a plane can be fully described by a set of three points in the two dimensional Euclidean space. Thus, the concept of spin falls outside of the Newtonian mechanics.

The word spin and process of spinning are used in everyday activities—like spinning of cotton or wool in which material is drawn and twisted into threads; or gyroscopic spin of a top; or twirling spin of tornado and so on. These spin representations have symmetrical motion of turning or rotational symmetry in the material; and the number of symmetrical appearance of the object while turning it or observing with 360 degree revolution around it is known as spin number. Similarly, spin in physics is also considered as number of symmetry appearing in the rotational observation or in revolution of an object by 360 degrees—like the spinning of a planet.

Thus, to accommodate our day to day understanding of Newtonian mechanics of two dimensional Euclidean spaces and its current usage in physics we define spin as follows:

*Spin of an entity in a plane is a number of its symmetrical appearances in 360 degrees observation or the 360 degrees rotation of the entity*

Under the above definition, a three point connected entity lies in a plane and its spin properties can be discussed without having a true spin; but by limiting to a particle—to a point—it is not feasible to talk about its spinning as the concept of spin demands to have more than a point in two dimensional spaces. And for the Newtonian particle, a point has no length, no width, no height, no front and no back, and so during its motion around the sun or around the planets, we cannot discuss spin, as spin is not permitted through 360 degrees observation or by revolution. But by extending the particle—the point concept in the plane, or a disk denoting a particle of the Newtonian mechanics, like a moon—which we see as a disk—as shown in Figure 4.2A. Now, consider a motion of going around a disk by 360 degrees from top to the bottom, or from side to side, or that of the moon going around a planet, earth, that does not change facing towards the earth while completing the orbit of 360 degrees. The symmetry does not change during the revolution or in the observation. Thus, the disk has 0 spin. This is the case of moon going around the earth. Under the 360 degrees revolution, we see the face—the front only and not its back—of the moon. Thus the moon has 0 spin. The moon does not spin as presented in Chapter I and shown in Figure 4.2A. We considered 360 degrees revolution of the moon as the entire disk—the full moon—is visible when it completes the orbit, during this orbit we do not see its back side.

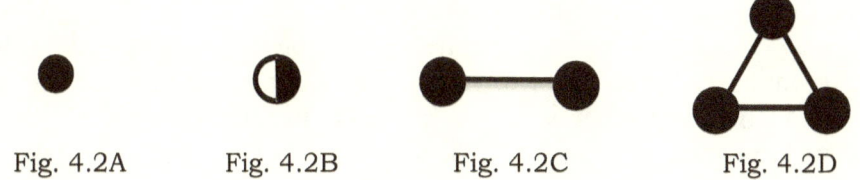

Fig. 4.2A          Fig. 4.2B          Fig. 4.2C          Fig. 4.2D

This is on a plane, by extending a line from the point of the disk or sphere, as shown in Figure 4.2B, and we will name it disk-line. We can revolve or observe the disk-line by 360 degrees. In this process, the disk-line will alter but at the end of the process it will return to its original shape. Thus, the disk-line has spin 1. This is also the case for the earth. We can stand on earth at a point facing the sun while earth revolves by 360 degrees we observe the light from the sun and darkness at night. Thus, the earth revolves by 360 degrees while it rotates around the sun. If this was not the case, we will have light on one half and darkness on the other half of the earth for all the time. But that is not the case, and so the earth has spin 1 as shown in Figure 4.2B.

In a similar extension of the disk-line, we can create, respectively, 2 and 3 spin entities by putting a second disk at the end of the disk-line point, and constructing an equilateral triangle with three disks on each vertex point as shown in Figures 4.2C and 4.2D. These are three examples of possible spins in a plane with three disks, or its equivalent structures of Euclidean geometry.

## 4.4 Spins of electrons and other objects in 3 dimensional spaces

The above presentation of spin is limited to three points connected entities residing on a plane. In reality the earth spins with four points connected gravitational forces created by sun and moon, which we presented in above discussion as a limiting case of rotation in a plane. The earth, planets, motions of electrons with magnetic fields, and spatial-entities in three dimensional Euclidean geometry required of having an extension of the definition of spin to accommodate other possibilities of rotations (spins).

In particular, Faraday introduced the concept of tubes (magnetic lines) of force require further clarification. The density of magnetic lines of force is known as a magnetic field, or magnetic flux density representing its strength. The magnetic field has magnitude, and a direction like any other vector force-field. In addition to these two vector properties, magnetic field density represents strength of rotation field in the space. When a magnetic needle is brought near a magnet in the space, the needle rotates (turns) according to the magnetic field strength in the space created by a magnet in the neighborhood. Maxwell worked with magnetic rotation, its induction, and established its relation with electrical charges in motion and expressed the relations in mathematical equations, known as Maxwell's equations which are simplified (and presented in Chapter V). Rotation of a vector $A$ is defined in Stokes' theorem, Chapter III. Based on these developments and simplified mathematical equations the magnetic field $H$ at a point in the neighborhood is given by

$$H = \nabla \times A = \text{rotation of } A \qquad (4.4.1)$$
$$\nabla . H = 0 \qquad (4.4.2)$$

where $A$ is another vector force-field originated by the magnet, and $H$ satisfies the condition (4.4.2), that the divergence of $H$ for all points in the space (neighborhood) is zero.

Equation (4.4.1) reveals two things. First, according to mathematical interpretation, the right-hand side of equation with rotation of a vector field $A$ produces another vector which is perpendicular to the surface in which $A$ lies, and equation represents a three dimensional picture for spin. If the vector field $A$ is conservative force—like gravitational force of attraction between two bodies, like that of moon and earth, or derived from a potential originated by the position of a body or an electrical charge—then the rotation of the vector $A$ is zero. But for the field represented by Equation (4.4.1), the rotation of the vector field $A$, the spin is non-zero. For example, the earth has combined gravitational effects from sun and moon, which produces a non-zero spin, which we discussed in Chapter I for the spin of earth. The earth spins around its axis which is perpendicular to the plan of gravitational fields from sun and moon. The spin vector associated with earth does not need to satisfy Equation (4.4.2) as it does not denote a magnetic field, and has only one direction of orientation to spin with respect to the vector field $A$. Similarly, if the vector field $A$ is non-conservative, and produces a rotational motion—like a vortex or motion of a top—then rotation of vector field $A$ is perpendicular to the plane in which vector field $A$ lies. In this case our earlier definition of spin holds good and earth has spin 1. But these cases present a three dimensional spin picture with a vector field is perpendicular to the plane (surface).

Second, the left-hand side of the Equation (4.4.1) for the magnetic field has constraint of Equation (4.4.2). A motion of electrons produces the magnetic field and establishes connection with space, as noted in Chapter II. We are unjust to electrons in considering as particles and waves, or particle/wave duality. In reality, once we relinquish the restrictions of Euclidean and Riemannian spaces in our study, electrons will reveal their true and much larger characteristic representations in space than that being portrayed by particles and waves. As a fact, with particle and wave limitations, electrons maintained their physical characteristics in three dimensional geometrical representations through the Equations (4.4.1) and (4.4.2), and other Maxwell's equations for the Euclidean space.

The Equations (4.4.1) and (4.4.2) imply that the rotation— spin— is a three dimensional entity, with a spin vector perpendicular to a minimum of three points connected surface, of which plane is a special case with zero curvature and torsion. From Equation (4.4.2), it follows that in maintaining the divergence of magnetic field lines to be zero at a point, the total number of magnetic field lines incoming to the volume should be equal to outgoing from the volume. Here the volume size is not defined, it is arbitrary. This is feasible only if the magnetic field lines at a point—under the discussion—have neither source nor sink to originate them, have neither beginning nor the end, and so must be closed loops.

Maxwell unified the electrical charges (electrons) with their electrical and magnetic fields in the neighborhood. The electrical charges have minimum of four points of connection originating spin produces magnetic field when in motion. The electrons behave like little magnets in orbital motion around the nucleus and also in free motion in space. The moving electrons originate current, and its associated spin acting as magnet originate magnetic field around the current.

The motion of electrons produces the magnetic field with spin-vector *H*, have two possible directional curves that originate from the point under consideration and are perpendicular to the vector *A*. From the logical understanding of Euclidean geometry and Newton's third law of motion, we can comment the followings on the vector-field *A* and magnetic field *H*.

1.  An electron has four points of connection in space, and in its motion adds extra points of connection, reveals it characteristics, particularly in its spin, how it is associated with the space, time and matter. The spin of electron is different from the one presented in Section 4.3 above and its characteristics require developing it step by step and geometrically representing it to reveal its structural characteristics.

2.  As discussed before, a (gravitational) material particle has minimum of two points, one each, material and geometrical point. An electron has two separate geometrical points (in addition to the material point). To identify these two geometrical points, let us first name or identify one of them as Front Observable (FO) point. The electron has front and back, or top and bottom or left and right sides. With respect to top and bottom, or left and right sides may be partially

observable—depending upon how it is observed, but the Back side is locally not observable. We will call it Non-Observable Back-side, and abbreviate as NOB. By locally observable and non-observable back we mean that when FO is observable, NOB is not observable, and vice-versa.

3.  During the spin of electron, the vector field *A* and magnetic field *H*, will reach to the non-observable point, which we have named and identify as NOB.

We start with electron as minimum of two point's material, with FO and NOB and have thickness T. This is a fundamental difference from the gravitational particle. We can consider the FO and NOB along the electric filed of the electron. The electron's motion adds a third point of connection to it producing the vector field, *A*, and the fourth point is due to the rise of the magnetic field *H*.

4.  To visualize our discussion and to portrait an observable picture, let us consider that the vector field *A* associated with motion of electron producing the magnetic field is on a closed loop orbit, as noted above, associated with the electron at point FO. When the electron completes the orbit along the closed loop, the vector *A* should not (and will not) meet at the same point FO of electron. If it is, then under the Newton's third law, the vector field *A* is equal and opposite and the rotation will be zero, which is not the case. Thus, the closed loop returns back to the point NOB of the electron, which was not observable before. These two points, FO and NOB, and the vector field *A* form a surface S that must accommodate one of the points of the magnetic field *H*, which is perpendicular the surface S.

5.  After completing the first closed loop revolution of 360 degree motion in the space, the vector-field *A* returns to the second point NOB of the electron. The vector field *A* has changed its orientation by 180 degrees (which we will describe in Item 8 below and reveal in the Figures 4.3 and 4.4), and it is in the opposite direction of the vector field *A* from which it started.

6.  Due to the electron's motion, the originated magnetic rotational vector field *H* is on the closed surface formed by the electrons closed loop orbit. The originated closed surface has the same thickness T, the material thickness of the electron. The vector field *H* is on the surface

and is perpendicular to the electric field and to the motion of vector-field *A*.

7. After completing the second closed loop revolution of a 360 degree motion in the space, the vector field *A* returns back to the first point FO of the material of the electron from where it started. The vector field *A* has the same orientation from which it started.

8. One can easily visualize the motion, the vector field *A*, the magnetic field *H*, and the electrical field by considering these entities on the Mobius band, as shown in Figures 4.3 and 4.4.

9. The Mobius band configuration is a well-known in topology. An individual unfamiliar with the configuration of Mobius band, can construct it by taking a strip of long paper and twist the strip by 180 degrees first and then turning it by 360 degrees so the two opposite back corners meet with the front two corners of the strip, and joining them with clear tape, as shown in the Figure 4.3 or 4.4.

 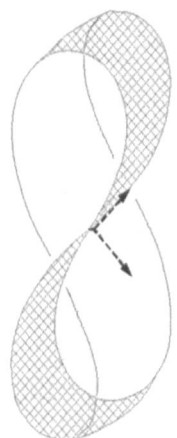

Figure 4.3            Figure 4.4

Thus the electron has spin and it returns back to its original starting orientation after having the 2 closed loops orbital motion of 360 degrees each, making revolution of 720 degrees.

From the above noted observations, to accommodate the current usage of spin in quantum mechanics and to accommodate and extend the presentation of Section 4.3 discussion, we define the spin of an entity in three dimensions as follows:

*Spin of an entity along an axis is a ratio of 360 degrees to the one with non-zero degree of revolution required to return to its original state.*

Based on the above definition of spin, one can verify that the Figures 4.2 B, 4.2 C, and 4.2 D, respectively, have 1, 2, and 3 spins, while the electron has half (½) spin. For the Figure 4.2 A, the entity remains the same with zero degree of rotation—to all including 360 degrees of rotation; thus no revolution is required, and so it has 0 spin.

10. In this twist of 180 degrees, one can twist it either on the right or on the left, giving two possibilities of the Mobius bands. We will work with right hand Mobius band.

Currently in study of electricity and magnetism, there are north and south poles for the magnetic field, and left and right hands of magnetic induction laws. We will focus on the right hand magnetic field law, and electrons as it is easy to represent geometrically and show it with an experiment having (1/2) fractional spin that are different than the integer number spins.

In this geometrical representation of spin, one can alternate the selection between the directions of motion of two geometrical points at a small distance, with locations of two electrons at a small distance on a circle, without any consequences to the presented discussion related to the magnetic field *H*.

To observe the discussion on Mobius band, let us consider a small plane on a configuration—the structure of Mobius band—with two vectors denoting the electric field perpendicular to the edges of the plane of Mobius band, and perpendicular to it is in the direction of the magnetic field *H*, the spin direction, and the direction of motion of electron is perpendicular to the plane formed by *A* and *H*, as shown in Figure 4.3. These three directions form a local triad of perpendicular lines to each other and are located at a point of interest. If we move around the triad of lines on the Mobius band by 360 degrees and return to the starting point, shown in Figure 4.4, we will observe that the vector **H** is equal in magnitude but opposite in direction, similar to the other perpendicular vector to the plane formed by electric field and the direction of motion, while other two vectors remain in the same directions. After having a second rotation around the Mobius band by 360 degrees and returning to the starting point, all the three vectors will be along their original directions and magnitude. In this case the symmetry with all the vectors and its orientation return back to the original condition (state) after a 720 degree rotation.

From the above representation of the magnetic field originated from the motion of electrons, one notices that it is hard to observe the two points associated with the material of electron; from the empirically observed facts, at the start of the motion—i.e., electron at rest has maximum electrical field. Now when the electron has completed one turn on its orbit, its material has maximum velocity and maximum magnetic field H, but has zero electrical field as it was presented by Maxwell. In this case, electron has the maximum momentum and is located at NOB. Thus, the electron in motion is observable at FO or NOB, but not both simultaneously, as it has a half (½) spin and we can only see either its front (FO) or back (NOB), but not both.

## 4.5    Observing a Magnetic field producing a couple and spin of electrons

The forgoing discussion has revealed three physical properties related to electrons and its associated fields; its applications to physics diverted us from their geometrical understanding and representation, and openly defied the Newtonian mechanics. We have paid a high price for its (electrons) defiance since its appearance in our day to day life—as it has not permitted in any earlier derivations from its properties to deviate to accommodate Newtonian enablers. The followers of electrodynamics did not know these facts, but have recognized it at the various stages of the subject development including the Maxwell's equations. And now Newtonian (and Quantum Mechanics) followers need to recognize and accept it—to sense that they are on the reserved realm—and have to change to accept the truth of the physical reality associated with electrons.

First, as noted in Section 4.4 above, the geometrical connections of electrons in motion having half spin can be satisfied only when the electron's electromagnetic—electric and magnetic—fields and its motion lie on an unoriented (Mobius) surface.

As we observe, two long-term Newtonian facts were used in understanding the motion of electrons that did not permit to realize its (electron's) kinematical unoriented property associated with its motion. The first involved the particle property associated with the material studied under motion. The Newtonian material particle is treated as a point (even it is a large body), has no orientation, and cannot accommodate any length, width, surface, nor volume associated with it. The forces of attraction between two particles act at a distance through the associated material points; and per Newton's third law of motion, action and reactions between materials are equal and opposite and are along a straight line.

By relinquishing both of the Newtonian properties noted above, the material of electron has four points of connection permitting to have point, length, width, surface, and volume with electrical field in its neighborhood. Its motion originates the electromagnetic fields. All these properties connect—locally as well as globally—with the surrounding space; and modifying the Newton's third law of motion, with the electromagnetic forces appearing at a finite distance—equal to the (minimum of) material thickness of the electron—on the opposite sides of the local surface associated with the electron. As presented in Section 4.4, the electromagnetic forces, which had originated due to the electron's motion, are equal and opposite, but are on the opposite sides of the surface associated with the electron. This fact of equal and opposite electromagnetic forces at a distance, similar to the third law of Newton, could not be visualized as long as we are within the conceptual realm of Newtonian mechanics. The Mobius band reveals the constraints associated with Newtonian mechanics, and establishes the local and global spatial connections of the force fields.

Second, contrary to the gravitational field, a magnetic field produces a couple, and not a force. To observe the magnetic couple, please perform an experiment described in Appendix A.

The benefit of most advantageous (opposite to venomous) part of the magnetic couple has seized the control of the couple appearing in Newtonian mechanics. The divergence of the magnetic couple over a given volume is always zero. This fact in general is not true for a couple of Newtonian mechanics and thus the magnetic couple meets the Newton's third law of forces by reducing the volume to zero.

To illustrate the point: one can spin a body (equivalent to a Newtonian particle) with equal and opposite forces originated from the field developed by a second body under the Newtonian laws, but will not influence other body lying in the field of two bodies. This is also the case for the magnetic couple, but it is with an exception. All material elements having electrons with connections of a particle, or a curve, or a body will not spin in the magnetic field. Only the material with surface connection with unpaired electrons located at a finite distance, like a paper clip made from iron, cobalt, nickel, or other similar materials will spin in the couple of magnetic field.

Third, materials, like iron, cobalt, nickel, with unpaired half spin electrons, when moved perpendicularly to the magnetic field spins, rotates around the direction of motion as an axis of rotation.

Electrons are hard to understand in their characteristics and motions. But electrons are canny enigmatic, unlike any we have seen in material particles of Newtonian (Quantum) mechanics. The particle electrons of Newtonian (and quantum) mechanics have yet not successfully explained their motions in the various atomic structures of periodic table.

For an irrational electron there is no simple straight line or a simple curved motion like the one presented by the particle studied in the Newtonian (Quantum) mechanics having a simple straight line or an elliptic curved motion. And as the electron moves from a FO point to have an orbital motion, it produces magnetic field in its orbital when returns to NOB point. In this case, the momentum is not at a point, but puzzlingly in the surface which is not an easy to observable at the NOB point

In Chapter V, based on various manifold connections of electrons we will see the complete structure all elements of the periodic table. For now, let us just say that the electrons with the surface manifolds are in motion in the (10) atomic elements falling in the fourth to seventh rows, and between second and third columns. The sixth, seventh, and eighth atomic elements of the 4th row—iron, cobalt, and nickel—are with unpaired surface electrons (and the same way the 5th row has similar elements). These unpaired surface structured electrons with half spin acts like spinning ball bearing and engage rotating when moved up down in the magnetic field. To observe the half spin rotation of these electrons follow the same method of these materials as can be done for the paper clip motion presented as an additional properties discussed in Appendix A.

The concept of a point particle panders in shameless fashion in the development of many physical theories other than Newtonian (and Quantum) mechanics. The developers of the theory of electrons motions are aware of it and in the beginning take that shape at the bases of the development. But from the above discussed facts, we need to be cognizant that the electrons have properties which cannot be accommodated by the concept of particles. These facts are presented in the above noted properties of electrons and are observed by simple experiments described in Appendix A.

## 4.6  Summary

From the above representation of the magnetic field originated from the motion of electrons, one notices that it is hard to observe the two points associated with the material of electron. And from the empirically observed facts, at the start of the motion—i.e., electron at rest has maximum electrical field—one can easily find its position with respect to a reference frame, let's say K. Now when the electron has completed one turn on the orbit, its material has maximum velocity and maximum magnetic field $H$, but zero electrical field as it was presented by Maxwell. In this case, one can observe the maximum momentum of the electron at NOB under the same reference frame K, but not sure about its position located at NOB

In the stock market there is a saying: "Men who can be both right and sit tight are uncommon." In the same way, by living on earth, and having a reference frame K associated with the earth, "it is uncommon to observe both the darkness of night and sunshine of the daylight," as the earth spins on its own axis and one can only observe either the darkness or the daylight depending whether the earth is facing away or facing the sun. And at dawn or at dusk, one is not sure about the clear daylight or the darkness of night, but only sure about the transition.

And for electrons, it is uncommon to observe the both—its position and its momentum—at the same time, as they both are on the opposite sides of its unoriented manifold. Consequently, it is not feasible to observe and measure simultaneously the position and momentum of the electron in motion.

It is hard to comprehend characteristics of enigmatic electrons. Those who have studied its properties carefully and followed its implications logically have brought developments of good results to physics and created family feuds among the physicists at each stage that revel its new property which is different from the contemporarily accepted properties by the founding fathers of quantum mechanics, who are not willing to change from their thought (of mathematical) position in working with the least action principle.

Here we see that the concept of connection clarifies the cause of the scientific community family feud—the uncertainty principle appearing in quantum mechanics—without addressing quantum mechanics and to the least action principle of quantum mechanics; simply by extending the territory of reign of classical theory of Maxwell's electrodynamics. There lies the strength of the connections introduced in this book.

# Chapter V

# The Rein of Electron Connections Removes the chaos in representations of Atomic elements in Periodic Table

## Why the disorder?

Because there are two things in science related to Space, Time, and Matter (STM)
—STM is macroscopically continuous, and microscopically discontinuous.

## How is it resolved?

Geometrical connection of electrons clarifies the chaos of atomic elements.
And it discloses electrons' shapes associated with Coulomb's particles, Stokes' curves, Faraday's surfaces and Maxwell's bodies; explains of having 7 rows and 8 columns arrangement of atomic elements in the periodic table.

## 5.1    Introduction

When I think about what it is to get through the motions of electrons in the atomic elements of the Periodic Table, I am reminded of a classic scene of old western movies in which a group of cowboys riding the horses travelling through an unknown, and possible hostile landscape. For example, Heinrich Hertz who died in 1894 at the young age of 37 while he unknowingly exposed himself for a long time to the hostile environment of radiation from high voltage discharges of electrically charged particles while performing experiments to demonstrate that Maxwell's electromagnetic fields represent the waves. Thus, these cowboys were not ordinary; they were all wranglers in their times, as it was bestowed upon Maxwell—the founding father of electrodynamics—as the 2nd wrangler in his tri-post exam at Cambridge University. These wrangler-cowboys did not know exactly where they are going in understanding physics of electrons and other related subjects, or what they will encounter in the development of their theory; they only knew that they would like not to turn back on Newtonian particle physics but must move forward with the faith and trust in themselves that they will eventually reach a place where things will be safer, better, and clearer than where they started at. This was the case for all of them. Hertz believed in the electromagnetic (wave) theory, which was theoretically developed by Maxwell and was sure it represented a wave theory—while his contemporary believed in particle theory—detected and physically demonstrated first time the electromagnetic waves under very high voltage. It is due to the sacrifice of Hertz and his trust in Maxwell's electromagnetic wave theory that we are enjoying radio, TV, wireless phones, and many more high tech gadgets today.

Similarly, going through the development of atomic elements of periodic table was a march through unknown territory. A simple understanding of an atom and atomic elements were unknown. The wranglers portrayed partial picture of atoms, atomic elements without direct participation of geometrical connections of electrons. In the following, we will summarize how the word atom came about, and how the atomic elements were organized in the form of a periodic table.

The Greek philosophers Leucippus, Democritus, Epicurus, Thales, and others used the word "atom" in the sense of being, each permanent and indivisible, that cannot be divided into many. This concept of atom lasted for almost for two thousand years. In 1661, Robert William Boyle (1627–1691) defined "an element" as a substance that cannot be broken down into simpler substance by a chemical reaction. These definitions served for a very long time in science; however, none of them are valid any longer after the recent observations of the subatomic particles. But at present, the concepts of atoms or particles are still in the ascendancy.

The "periodic table of atomic elements" is about two hundred years old and reflects its growth in discovering and understanding the chemical and physical properties of the atomic elements. The development of the table recognized patterns and properties of the elements and modified its shape and arrangement. In the beginning only a few elements were known, and Henning Brand (1630–1710), William Odling (1829–1921), John Newlands (1837–1898), Julius Lothar Meyer (1830–1895), William Ramsey (1852–1916), and others participated in finding more new elements.

The most important events related to the periodic table, respectively, occurred in 1869 and 1870 when Dmitri Ivanovich Mendeleev (1834–1907) and Julius Lothar Meyer independently published the tables of atomic elements. Mendeleev arranged elements in vertical columns, named as groups, according to increasing atomic weights, and exhibited apparent analogues periodicity of their physical and chemical properties in horizontal rows, named as periods, and proposed that the gaps represented elements that are yet to be discovered. The Mendeleev atomic table recognized the relations between the properties of the elements and their atomic weight (mass), and the subsequent discovery of the new elements matched with the predicted properties helped to acceptance of his table in science.

For the periodic table, Mendeleev discerned the regularity of chemical properties of the elements according to their atomic weight (their masses). In the arrangement of the elements, he discovered that elements with similar chemical properties occurred periodically and could arrange them into definite families in the table. The presented atomic table appeared depending on the atomic weight (mass) of the elements.

For the periodic table, Mendeleev did not predict the existence of a family of elements of the noble gases, which was invented by William Ramsey and added at the end of the groups. According to the Niels Bohr (1885–1962), the model of structure of atom of Ernest Rutherford (1871-1937) (which we will discuss further in this chapter), weight (mass) of the atom is not that important as far as its chemical behavior is concerned, but rather the charge on the nucleus—that is, the atomic number—determined by the number of protons in the nucleus. In an atom, protons and neutrons appear in the nucleus, and they are collectively called nucleus. Per current accepted terminology of proton-neutron model of nucleus, the number of protons in nucleus is the atomic number of the element in the periodic table. The number of protons plus neutrons is the element's atomic mass number, as the mass of the orbiting electrons, around the nucleus is negligible. In the atomic table, as one moves from the one element to the next higher element, the number of protons increases by one so that the charge on the nucleus, and hence the atomic number increases by one.

For the current representation of periodic table, the work of Henry Gwyn J. Moseley (1887–1915) showed that the atomic number, and not the atomic weight, is decisive in quantity in the arrangement of the elements. He showed that the chemical elements can be arranged in a step by step sequence, by adding one unit of charge at a time, carries the implication that numerical sequence is not broken, and no unsuspected element can be present outside and could be found in the periodic table. Moseley included all possible elements in the atomic table including the noble gases. Thus Moseley's presented periodic table is complete with all elements by adding positive charges of protons one at a time to the nucleus starting from the first element—hydrogen.

## 5.2 Preparations to simplify the atomic table

Moseley investigated the spectral lines emitted by the heavy elements in the X-ray part of the electromagnetic spectrum and stated the relationship for the elements in a simple empirical formula that match with the experimental data. The table has stood the test of time. The researchers working in quantum mechanics have explained various properties of the elements with long theoretical clarifications based on the spectrum of light observed under various physical conditions, without providing explanations on the simple and fundamental factual question of the periodic table: *Why are there 7 (period) rows and 8 (group) columns in the periodic table?*

According to Robert Andrews Milliken's (1868–1953) work, a fraction of electron does not exist, but fractions of electric charges for protons and neutrons—each expressed in terms of fractions of electrical charges as sum of (three) quarks—do exist. So we will focus on the electrons and their associated number of points of connections in developing and understanding the rows (periods) and columns (groups) associated with the periodic table.

As observed through experiments, most of the chemical elements making up the material world have isotopic form. Frederick Soddy (1877–1956) and E. Rutherford showed that in the radioactive process the atoms of one chemical element could be transferred into the atoms of another chemical element. This radioactive transformation could give rise to the isotopes of the atoms that differ in their atomic weights, but are chemically equivalent with the original element. An isotope of an element has the same number of electrons and protons, but has a higher number of neutrons, and thus a higher atomic weight. Due to these observed facts, the observed weights of the chemical elements are determined by their isotopic composition and require having an additional spectroscopic investigation to observe the elements' (nucleus) structural details. But, there is no change in the number of electrons for the elements and their isotopes and they orbit in the same orbits. So by focusing on the number of electrons and their connections permit us to study the atomic structures, and well defined their characteristics of all elements in the periodic table.

The above noted basic differences, together with unsettled issues of the particle and wave nature associated with their matter, suggest that electrons possess properties larger than corresponding to the idea of a simple particle appearing in Newtonian mechanics, and waves in Hertz's presentation of Maxwell's electrodynamics. They suggest that the speck of any element of a material appearing in the space and time associated with the electrons is not limited to be a simple (material or geometrical) point, but it has additional properties.

To reveal these properties, we raise these material points of electrons to a higher status through a postulate:

*Electrons are manifolds that permit them to unite with the space, time, and matter.*

As we have discussed in Chapter IV, the electrons have a half (½) spin and we can observe only the front or non-observable back side at a given time. So, we introduce a second postulate associated with the manifolds in the atomic elements, namely:

*Electron manifolds have local properties of having front and non-observable back-side (obverse) (associated with points), or directions (associated with lines or curves), or top and bottom (associated with surfaces), or inside and outside (associated with volumes).*

These geometrical properties neither explicitly influence nor implicitly require, but tacitly participate in the studies of Maxwell electrodynamics. These manifolds change their shapes and sizes according to the neighboring space and time, influence from other charges and circumstances of their orbital motions in atomic elements.

It follows from Newtonian mechanics that the four points connected particle (a planet, like earth) orbits in a closed elliptic orbit and permit to have a spin. The four points connected manifold structures of electrons with Newtonian time and steady state elliptic motions around the nucleus are adequate to derive the structural details of all atomic elements of the periodic table. It is the extra points of connections available to electrons in their motions appearing in the observations of the atomic elements and so with the manifolds.

In this chapter we will focus the discussion on the maximum of four points connected manifolds of electrons. The electrons have higher than four points of connections possible when they are moving on and out of the elliptic orbits and under higher energy than the one permitted by the steady state motions around the nucleus. In developing the periodic table of atomic elements we will not consider the case in which electrons move out of the orbits or out of their shells, as discussed in quantum mechanics.

## 5.3 Manifolds in electrostatic, magnetostatic and electrodynamics

Electrons manifolds have appeared in various forms in electrostatic, in magnetostatic and in Maxwell electrodynamics. The physical studies either ignore or represent the manifolds in terms of other geometrical and physical properties. We will reveal how these manifolds have emerged in the past as special cases in the electrodynamics of Maxwell's equations:

$$\nabla . E \quad = \quad \frac{\rho}{\epsilon_0} \tag{5.3.1}$$

$$C \nabla \times E \quad = - \frac{\partial H}{\partial t} \tag{5.3.2}$$

$$\nabla . H \quad = \quad 0 \tag{5.3.3}$$

$$C \nabla \times H \quad = \quad \frac{\partial E}{\partial t} + \frac{I}{\epsilon_0} \tag{5.3.4}$$

In this chapter we will denote all vector fields with italic bold letters. In an electrostatic case of Coulomb's law between two charges at rest, electrons appear as particles, and the force between the two appears as the product of their charges and inversely proportional to the square of the distance between the two points where the charges are located. The Coulomb's force converts into an electrostatic field with electrons appearing on a surface S such that its curl is zero per Equation (5.3.2). In this case, it has a given divergence and no effect of time dependent magnetic field on electrostatic field. The electrostatic field appears as three dimensional entities on a surface S. The field appears on, and not inside, the closed surface S. The four points connected electrons represent as the one having three points connected surface with electrostatic field *E* and the fourth point is at the location of the charge.

A magnetostatic case is an approximation and it is only possible when there is a steady flow of fixed amount of charges—electrons—in the space. In this case, the magnetic field (*H*) lines and field surfaces are examples of zero divergence by Equation (5.3.3). The divergence of a field is zero for any (size) volume. Depending on the volume without charge or with charge, respectively, the divergence of the field *H* is zero for a point (with no charge), or a closed surface (with a charge). From these observations, it follows that the manifold of electrons is such that its circulation is zero at a point if there is no charge, or it produces only on a closed surface with two or more charges apart and it is such that its divergence is zero by Equations (5.3.3). Thus, the magnetostatic field *H* follows from a curl of a vector field *B*. In this case, the manifolds of electrons have eight point spatial-connections, produced by a minimum of two four-point connected electrons moving under the steady-state motion.

From the above noted electrostatic and magnetostatic cases, one can introduce a time dependent motion of electrons—currents and magnetic fields—and derive Maxwell's equations of electrodynamics. The Maxwell's equations are independent of north and south poles of magnetic fields, and the positive and negative charges of electrons. The fields and the currents depend only on its neighboring space, and their changes with the space and time, and need to satisfy the Maxwell's equations.

There are eight Maxwell's equations for electrodynamics and eight points of connection for freely moving electrons. Time participates in establishing these connections of the associated fields. These eight equations uniquely and completely define the motions, and provide the required eight points of spatial-connections for the electrons. Neither the Maxwell's equations nor the spatial-connections impose any restriction on the details of the surrounding fields or the neighborhood in which the (macroscopic structured—the minimum of four points of spatial-connected) electrons motions take place.

The above noted facts demonstrate that the electron manifolds change their characteristics of a spatial-connection of a point to eight points, change their shapes and sizes, starting from a point, to a curve (line), to a surface (plane as well as wave), and furthermore to a volume according to their motions in their neighboring space. We reported in Chapter II that the space and matter are united. The time permits the motions of these united entities associated with the electrons, magnets, and their fields. The time does neither necessarily require uniting with the space nor with the matter for the Maxwell's electrodynamics. In the derivation for Maxwell's equations, time remains absolute (Newtonian) and does not require to be relativistic.

So, for the study of the periodic table, we will consider the time to be absolute, and will focus to study the atomic elements under steady state motions of electrons as there is no change in the electrons, electric and magnetic fields and the associated currents. Similarly, there is no reason to develop the structure of elements from the light or X-ray spectrum observations, as the structure follows from the Maxwell's equations.

In using the various manifolds, however, we must be clear about what types of manifolds—the point, curve, surface, or volume manifolds—originated from the motions of electrons. The material of an electron occupies a point in space which we will call a material point. In addition to the material point, there are two types of points linked with electrons material points in their motions. These points are associated and spatial points. An associated point is directly linked with the material of electron, and it is there before any motion of electron occurs. A spatial point or derived point is a local one and located in the neighborhood of the electron's material and also appears in the fields originated by the electron motions. The spatial point can substitute with the associated point in the construction of manifolds. In motion, electrons can interchange the associated point with the spatial point without creating any physical changes. These three types of points—material point, spatial point and associated point—generate various geometrical and physical structures: points, curves, surfaces, and volumes that create various manifolds of electrons for atomic elements.

The electrons manifolds, however, do not permit to have multiple associations from the same field (associated points) or the domain (spatial points). One can have a geometrical configuration from these available points—a curve, a surface, or a volume—built with multiple points from the same field or from the same domain that will not produce the real physical manifold characteristics of the electrons. We will encounter examples of geometrical volumes in Subsection 5.5.4 that are not true (physical) manifolds, and we will not use them in the development of the periodic table.

## 5.4 Reasons for the 7 rows (periods) and 8 columns (groups) in the periodic table

According to the structural model of atom presented by J. J. Thomson and its further developments, it consists of a nucleus of a system of particles positively charged protons and neutrally charged neutrons, surrounded by a system of electrons kept together by attractive forces from the nucleus. In a normal ground, the unexcited and stable state of an atomic element, the total number of negatively charged electrons is equal to the total number of positive charge of the nucleus. The nucleus carries the essential mass of the atom, and its dimensions are much smaller compared to the dimensions of the atom. The number of electrons in an atom is equal to the number of protons in the nucleus and it is approximately equal to half its atomic weight. For the isotopes of an element, there are higher numbers of neutrons than the number of protons in the nucleus, so they are heavier but we will not discuss this case of atomic element.

The model of Bohr, based on the conditions of quantum of action and energy of electron moving around the nucleus, provided very good information, matching with experimentally observed spectral lines of hydrogen atom, in which one electron moves in a form of planetary orbit around its nucleus. Its subsequent applications to nuclear physics will probably not be replaced by any other model. Bohr's model has done three things. First, it led to a description of a stable electronic orbit. Second, it permitted to have a force between the negatively charged electron and positively charged nucleus (proton), similar to the Newtonian force of attraction between two bodies, and so the energy. He used the quantum condition on the energy for explaining the discrete spectrum. The quantum condition restricted the possible energies with the possible orbits of the electron. Following Planck, he identified certain energies' states with its orbit numbers, called as variable quantum number and denoted by an integer number n (and the number n may correspond to the atom's shell number). For the larger energy states, like the hydrogen, the atom had bigger orbits. Then, Bohr applied the principle of conservation of energy, i.e., the spectrum lines of emitted light, by identifying the energies as states of electron moving from the higher energy state (shell) to the lower state (shell) of the hydrogen atom. And the most stable configuration for the hydrogen atom is the state with lowest energy, called the ground state. Third, with this reasoning, he explained the part of Johann Jakob Balmer's (1825–1898) series of observed discrete line spectrum for the hydrogen atom.

Bohr's model's extension, even to the next simplest atom of helium, lithium, having more than one electron around the nucleus was not possible. As a fact, if Bohr had focused on any atom other than hydrogen, he would have failed. Apart from these shortcomings of Bohr's model, the legacy of Bohr's atom framework is pervasive. In fact, based on the Bohr's model, many scientists had worked to improve the atomic structure and the quantum mechanics.

There are similarities in the atomic model presented by Rutherford and Bohr and that of the solar system developed by Tycho Brahe (1546–1601), Kepler, and Newton. In each there is a massive core that exerts a controlling influence over the less massive particles orbiting around the central core; the force between the core and the particles decreases along with the square of the distances. Bohr considered the orbit of an electron in hydrogen to be circular, similar to the case of the solar system considered in early seventeenth century. Note that an elliptic orbit has major axis and minor axis, whereas a circular orbit has both the axes equal to the radius of the circle.

Based on the observed exceptionally correct and accurate data for the solar system, Kepler, by abandoning the Brahe's idea of circular orbit, introduced the first law of the planetary motion, i.e., planets move in an elliptic orbit with sun at one of its foci of the ellipse. On the other hand, Bohr was not guided by the actual observed data from the orbital motion of the electron, but by the reasons of the simplicity to explain a part of the Balmer series of hydrogen spectral lines, which upon close observations included not one line, but two or more lines. So Bohr had a hard time to explain the spectral line observations with his circular orbit model.

To accommodate multiple lines in the hydrogen spectrum, Arnold Sommerfeld (1868–1951) generalized Bohr's model by considering the elliptic orbit, as Kepler did for the planets, and quantized the space with relativistic effect on the mass of an electron.

The elliptic orbit provided two integer variables to discuss the states of energy spectrum, one of which corresponds to the quantum number n based on the major axis of the ellipse, and the second variable corresponds to minor axis and denoted by k, varying from 1 to n. These two variables accommodate the spectrum lines for hydrogen and hydrogen-like atoms—helium, lithium and so on (which we will see in Section 5.5.4)—that correspond to the elements having electrons acting like particles.

The work of Bohr, Sommerfeld, and others provided enormous valuable information in the analysis of the atomic structure; other methods of representing atomic process have been in better agreement with the experimental facts. So, Schrödinger introduced a wave model for spectrum of light that greatly improved and extended the theory with the Legendre Polynomials expressing their states in terms of variables denoted as s, p d, f, g, and h, representing possible numbers of its state and developed the orbits. The results are good, but unclear on the shapes, size, and the physical characteristics of the electrons in their orbital motions around the nucleus. The subject is interesting and the reader is suggested to refer to books on quantum mechanics or on spectra of atomic elements.

At about the same time, Werner Karl Heisenberg (1901–1976) and Paul Adrien Maurice Dirac (1902–1984) focused on the observable quantities of quantum theory—namely on the observable frequencies and intensities of spectral transitions, and put forward matrix mechanics (quantum dynamics), as the coordinates of the electron and their orbits were not observable. Due to the observable facts, matrix mechanics also became successful in its general applications. While both these theories are different in their mathematical formulations, Schrödinger showed that they lead to the same results and in reality these theories are equivalent.

In spite of the equivalence of the two theories, the differences between Schrödinger's and Heisenberg's approaches originated criticism from one to the other's ideas of developing and applying to understand the physical structure of atoms, and overlooked the essential foundations on which their theories rest. I used the word "rest" rather than "based" because both the theories faced, time after time, the need to modify their foundations. In the beginning, the theories appeared promising, but when viewed on the ideas like the probability distribution, uncertainty in observation of position of electrons and their velocity, duplexity phenomena of the observed and theoretical considered electrons, etc., both theories fell short. These theories are excellent, but have yet not obtained the status similar to the classical methods of physics and mathematics. Both the theories used the spectrum of light, and did not address to the fact that the motion of electrons in steady state can be described in terms of the classical electrodynamics by taking into simple account the electrons' structural manifolds.

We will remain within the tenets of classical electrodynamics, without using any of the reasoning from Thomson or Bohr, or quantum mechanics, and will represent all the atomic elements of the periodic table without addressing to the spectrum of the light emitted by the atomic elements. We will develop the structures of all atomic elements in the following sections, but in this section we'll explain why there are 7 rows (periods) and 8 columns (groups) in the periodic table.

## Explanations for 7 rows (periods) in periodic table

The electrons' spatial-connections and their manifolds simplify to have geometrical configurations of the atomic elements. The four point spatial-connected materials have spins as demonstrated earlier for the planets and in particular that for the earth. Electrons have a minimum of four points of spatial-connections, empirically demonstrated to have spins and in motion up to eight points of spatial-connections. These are the fundamental properties of electrons, whether the electrons appear as a simple particle (point) or up to four points connected (body) manifolds.

The significance of observations on everyday objects, like the development of the periodic table, has shown far reaching effects beyond their apparent simplicity and innocent look. For example, electrons have a shape of a particle, a curve (line), a surface (plane), or a volume as shown in the Figure 5.1. This figure is another representation of electrons than the one shown in the Figures 4.1A, 4.1B, and 4.1C of Chapter IV. This is the simplest possible four-point connected material tetrahedron for an electron with curves (1, 2), (1, 3) and (1, 4). These curves are perpendicular to each other to form a tetrad at the material point 1. Shortly, we will see that this simple property of an electron has a significance of seven possible manifolds for atomic elements to maintain the connections of electrons producing 7 rows and no more in the periodic table.

A tetrahedron is a convex polyhedron with three congruent faces meeting at the four vertices. A Platonic solid is the one in which each edge of the faces is of equal length, each vertex has the same number of faces, and the angles of the faces at each vertex are equal as shown in the Figure 5.2. The tetrahedron for an electron is not a Platonic solid. However, they can both topologically transfer from a volume to a surface, to a curve, and to a point without destroying their geometrical structure connections. This transformation is continuous.

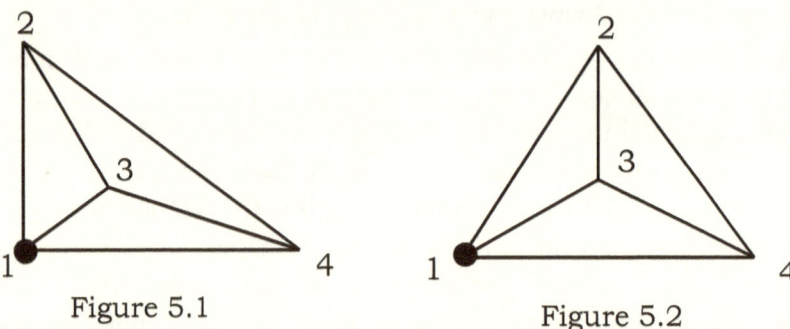

Figure 5.1                    Figure 5.2

   To associate the geometry of four-point connected manifolds to the physical properties of electrons, let us consider the material point of electron is at its gravitational and geometrical point (1). The three physical quantities for the electron—the material point, the gravitational point, and the geometrical location point—all of them are united at point (1). To distinguish it from other associated points with the electron, we will call it as a material point in the discussion. The electron has other associated points—that are the second, third, and fourth points, respectively, in the direction of the electrical field, magnetic field, and direction of motion in its orbit. These three curves (lines) are perpendicular to each other, as shown in the Figure 5.1 and are physically present in the motions of electrons.

   A (material) point has a front and a back (or top and bottom, or left and right sides), but has neither length, nor width, nor direction. A curve (line) in a space has a direction and its opposite direction. A surface (plane) has a top and a bottom. A volume has a front and a back (or has top and bottom, or left and right sides, or if hollow has an inside and an outside). These manifolds' physical and geometrical properties are observable with respect to the other physical entity (nucleus). Based on these properties, the point has a set of only two possible manifolds, each having a geometrical point property; the other three cases have two sets of manifolds, each having a specific geometrical orientation with respect to the other physical entity. (We will see the details in the following when we discuss the columns of the elements.) These seven possible manifolds associate with the material of an electron provide seven rows in the periodic table.

Corresponding to each set of manifolds there is one set of possible orbits for an atomic element. Thus, there are seven sets of possible orbits available for electrons to move to produce the maximum of seven rows in the periodic table of atomic elements.

**Explanations for 8 columns (groups) in periodic table**

In the history of periodic table of atomic elements, what can be considered as a historic event in chemistry is the introduction of arguments to have 7 rows (periods) and 8 columns (groups) associated with the atomic elements, during the time when the structure of the elements was unknown. At the time of the birth of periodic table, only a few elements were known. It was impossible for any intellectual person to assert a proposition of these columns (groups) for the elements, and these columns (groups) are still significantly accepted by the people purporting to care about the evidence. Thanks to the courage, amusing eccentricity to predict, and depth of understanding of atomic elements by people like Mendeleev, Meyer, and others to predict the periods and groups; Ramsey by predicting the noble gases to fall in the eighth column (group) based on inert properties of the gases for maintaining the nobility, nearly having no chemical reaction with other elements.

One of the basic properties of the electrons is that it returns back to the properties of particle under the classical Newtonian mechanics, which we noticed in Coulomb's law of attraction (and repulsion) between two electrical charges when electrons appear as particles and in Bohr's atomic mode of hydrogen where electron orbits like planets. After that the electrical charges (electrons) take off and reveal other properties that are different from Newtonian mechanics.

The first row of the periodic table has only 2 atomic elements—hydrogen and helium—whose electrons act as a particle. Hydrogen is the first atomic element having one electron, one proton and no neutron. In hydrogen, the electron acts like a particle and orbits like a planet, and have four points connection and spins while orbiting. After one complete revolution, the electron turns to its non-observable-back side (obverse), and returns back to the front side after a second revolution. Hydrogen appears in the first group of the periodic table.

Helium is the second atomic element having two electrons, two protons, and two neutrons. In Helium, two electrons act as particle manifolds. In helium atom, two electrons are orbiting on the same orbit at 180 degree apart. For stability and symmetry in the orbit, one of the electrons will be facing the front and the other will be facing the non-observable-back side (obverse). Helium is a noble element.

To keep up with the seven groups of the periodic table presented by Mendeleev, Ramsey called the detected noble gases residing in group 0. To recognize that the elements of the noble gases have all orbits occupied with 2 electrons each, we will call this group as column 8. By naming the noble gas elements group as the 8th column reminds us of the fact that the atomic periodic table has 8 columns, and electrons have 8 points of connections, and in particular, the noble gas elements are saturated with all permissible electrons in the orbits of that particular row.

For the first row, there are only 2 atomic elements that are possible with electrons appearing as particle manifolds. The helium atom orbit is saturated with possible electrons in the orbit, creating a stable and inert noble gas for the row, and falls in the 8th column of the periodic table. Helium is at the core of the elements in the second row to follow.

Similarly, in the second and third rows of elements, electrons act as particle and curve manifolds. As noted for the first row, there are only 2 elements possible with particle manifolds on a single orbit and 6 possible elements with curve manifolds on 3 orbits. These 8 elements create total of 8 columns. The last elements in the second and third rows, respectively, are neon and argon, and each of the four orbits is occupied with two electrons and is saturated. Neon and argon are noble gases and they fall in the eighth column of the periodic table. We will see in Sections 5.5.1 and 5.5.2 that in the eighth column, the atomic elements are fully saturated with two electrons in each and every possible orbit occupied creating noble gases of the column. These noble gases are at the core of other elements of the row to follow. This fact appears for the other noble gases; we will not repeat this comment.

In the first three rows, electrons create the eight columns (groups) for the periodic table by keeping their (classical Newtonian) particle and their naturally available linear (curve) manifolds, as well as their spin properties. The electrons in these columns neither depend on any external geometrical point available in their motion nor on their magnetic movement. These columns demonstrate that electrons appearing in the first 18 atomic elements are of true particle and curve manifolds, and were presented in the nineteenth century without addressing the spectrum of light. Thus there are only 8 possible columns for the atomic elements of the periodic table.

Recall that most of the elements occurring frequently in nature fall in the first three rows and their atomic weights are between 1 and 40. Other remaining abundance elements are in the earth's crust that are of less than 132 atomic weight and fall in the fourth and fifth rows. The rest and rear-earth and radioactive elements, are heavier and have atomic weights of over 132.

The atomic elements in the periodic table in the fourth to seventh rows have nucleus with more than eighteen protons and neutrons, and have at the best one electron acting as a particle. The atomic element with the particle manifold electron falls in the first column of the rows. Because of the spin of the particle electron, the elements with two particle manifold electrons in the orbit have to go together. So the second elements of the rows with two particle manifold electrons fall in the second column of the rows.

In Section 5.5.3 we have presented that the electrons in the fourth and fifth rows are in the order of particle, surface, and curve manifolds and, respectively, have 2, 10 and 6 atomic elements in each row. And in Section 5.5.4 we have presented that the electrons in the sixth and seventh rows, in the order of particle, volume, surface, and curve manifolds, respectively, have 2, 14, 10, and 6 atomic elements in each row. These experimentally observed structural facts related to the fourth to seventh rows rightly reveal, without addressing any theory of quantum mechanics, and convincingly portray the true geometrical picture of the positions of the volume and surface electrons elements in the rows. These facts follow simply from a simple picture appearing in the Euclidean geometry and properties of electrical and magnetic fields. Also, note that each row's first element starts with a particle manifold electron revealing the periodic property associated with the rows, and shows how it develops with physical structures of the electrons.

Before we move further, we must pronounce that, the members of first column elements—hydrogen, lithium, sodium, potassium, rubidium, cesium, and francium—are with one particle electron manifold in the orbit, and except hydrogen, create group 1 and are collectively called the alkali metals group. The alkali metal elements are chemically reactive as they have only one particle electron in the outermost shell. This particle electron can easily move out of the shell to other element in need of an electron in a chemical reaction. Thus, elements in group 1 often form singly charged, positively charged ions, such as sodium ion $Na^+$. The members of the second column elements—beryllium, magnesium calcium, strontium, barium, and radium—are with two particle electron manifolds in the orbit and create alkaline earth metals. The alkali earth metal elements are less chemically reactive as they have two particle electrons in the outermost shell. These particle electrons are orbiting in a stable state and so do not easily move out of the shell to other elements in a need of an electron in a chemical reaction.

While the members of the eighth column elements—helium, neon, argon, krypton, xenon, radon—are with all their available orbits filled with electrons, are inert and create group 8, and is called the noble gases group. The noble gas elements are saturated with required electrons in the orbits in the shell, and no electron would move out of the orbit to the other electrons in the neighborhood in a chemical reaction. These gases are at the core of the next row of the elements. On the other hand, the members of the seventh column elements— fluorine, chlorine, bromine, iodine, and astatine—are halogens and have almost all their orbits filled with possible electron manifolds with shy of a one curved electron manifold in the orbit to be full, and create a group 7 called halogen group. Due to the noted deficiency of electron in the orbit, similar to the alkali metals, the halogens are highly reactive. In chemical reactions halogens are ready to receive an extra electron to form a stable shell. For this reason, the halogens form negative ions to receive an electron, such as chloride ion $Cl^-$.

Out of eight columns, i.e., eight groups, the first two and the last two—thus four columns—are with dedicated types of atomic elements in the periodic table. Looking at the remainder of these groups, the elements in the third, fourth and fifth groups are with one electron in each orbit. These elements are ready to receive or release the electrons. While those in the sixth group are with two electrons in one orbit and other two are with one electron each. Due to these reasons, these elements are highly reactive. One can find more details on the chemical properties of these groups in most chemistry books.

Similarly, we will see in the following sections that the fourth and fifth rows of elements have 10 surface manifold electrons that fall between the second and third columns of the periodic table. And those in the sixth and seventh rows of elements have 14 volume manifolds, respectively, which are named as lanthanide and actinide series elements fall after the first element of surface manifold of the rows. We do not see either a physical or a mathematical reason for these 14 elements to fall between those columns, but after the first element of third column, as represented by the Moseley-prepared periodic table, rather than between second and third columns in order to follow the same logical development of the manifolds of the fourth and fifth rows. This offset of the atomic elements raise a question: Why did nature selected this order? These facts can only be decided by experimental evidence and if necessary, we may need to change the order of the Moseley's atomic elements periodic table.

## 5.5 Geometrical configuration of electrons in Atomic elements

Let us denote a point with no (zero) dimension, a curve (line) with one dimension, a surface (plane) with two dimensions, and a volume with three dimensions manifolds associated with electrons as $ne^{0i}$, $ne^{1j}$, $ne^{2k}$, $ne^{3m}$, where the first superscripts—0, 1, 2, and 3—respectively, denote the dimensions of the manifolds of the electrons and the second superscripts i, j, k, and m are non-negative integer variables, denoting the possible number of electron manifolds having connections with the nucleus, and their values which will follow from the discussion related to the motion of electrons' in the $n^{th}$ row, where n runs from 1 to 7 non-negative integers.

Let us recall that the electron manifolds are in a planetary orbital motion around the nucleus of an atom. In a steady-state motion, the manifolds confine themselves in their row, and avoid crossing from one shell to the other and maintain their stable structure.

For a stable structure of a specific row of atomic elements, electron manifolds starts occupying the possible orbits with point manifolds and end it with curve manifolds. Once all elements of a specific row get completely filled with all the possible number of electron manifolds, the last element of the row reaches to a noble state of the row.

In the noble state, the last element of the row has occupied all the possible manifolds of the electrons in all the orbits, and reaches to a saturated stable state. The noble element provides a stable ground state structure for that row and also for the next row of the elements. We will use the word 'shell' for all the elements of a specific row including the last noble element that permits to develop the elements of the next row. Based on these empirical properties of the atomic elements we will develop the periodic table.

To establish the number of possible orbits with electron manifolds and to fill the orbit, we will consider from the empirical facts that the specific type of manifolds occupy a possible number of orbits one by one. Once the specific type of manifolds has occupied the possible orbits with the noted specific type of manifold, then the manifolds will restart saturating the orbits with a second set of manifolds. The orbits can accommodate the maximum of two manifolds of the same type, same orientation but different spatial-connection. From this fact it follows that the derived number of possible orbits will be half of the number of possible manifolds of a specific type for a specific row.

The order of electron occupancy needs to follow the development of manifolds, and to provide the stability in the elements. The first row of the elements is the simplest—it has 2 point manifolds. These 2 point manifolds occupy the first orbit in each of the following rows as this one is the simplest of all manifolds. The second set of manifolds is the curve, and it occupies the second and third rows of the elements. These manifolds follow the point manifolds. The sets of point and curve manifolds have the four points of connection, and do not need any additional point to have the orbital motions around the nucleus. These two sets of manifolds—the point and the curve—provide the needed stability from the fourth to the seventh rows of atomic elements, and are, respectively, at the beginning and at the end of each row.

The third and the fourth sets of manifolds, respectively, are the surface and volume manifolds. We will see in the Subsections 5.5.3 and 5.5.4 that these manifolds, respectively, need extra spatial-point, and spatial and temporal points to have the orbital motions around the nucleus, though the time (temporal point) does not directly participate in the presentation of the manifolds and remains absolute— Newtonian—in the motion of electron manifolds. This added spatial point makes the surface manifolds unbalanced, and so their orbital motions fall between orbits of the point and curve manifolds. In the same way, the volume manifolds are more unbalanced than the surface manifolds, and their orbital motions fall after the point manifold orbits, but before the surface element orbits, and the curve elements remain in the last orbits for the fourth to the seventh rows of the atomic elements.

We will use the words—a line and a plane—respectively, for a curve of one dimension and surface of two dimensions either to clarify or to simplify the subject under the discussion.

## 5.5.1 Atomic elements of the first Row in Periodic Table

Recall that we are working with 3 dimensional geometrical spaces; the space, time and matter are absolute, and we will keep all of them (Newtonian) absolute. Also from the Euclidean geometry it follows that it is easy to define and detect a normal to a curve (line) or to a surface (plane), but not to a point or to a volume manifold. But for a normal to a point and to a volume manifolds it is not easy to define a normal to a manifold, so the normal has to follow from its connection and orientation with nucleus in the neighboring space. We will use these facts in developing the geometrical structures of atomic elements of the periodic table.

The first manifold of an electron is a point. In the first orbital motion, there is a point electron (particle) orbiting around the nucleus.

An electron is moving as a point manifold in a planetary orbit of the first element of the periodic table—hydrogen. This case is similar to a planet moving under a gravitational field that satisfies Newton's laws of motion. There is no reason for the electron to collide with the nucleus.

The first element, hydrogen (H), has a single electron that moves around its nucleus. The electron manifold spins in its orbit due to its four point connection (property), like the earth spins around the sun. The electron's negative charge balances out with the nucleus' positive charge. The hydrogen element is in its neutral field condition. We will denote the electron as $1e^{01}$ as it is in the first row with a point manifold with zero dimensions, as noted before. In this case the electron has four points of spatial-connection. First is at its gravitational point, second is towards the center of the nucleus, third is in the direction of its motion, and fourth is along its spinning.

Let us recall that the point has front and non-observable back (obverse) sides with respect to the nucleus. So there are two possible electrons satisfying the required eight points of spatial-connection in the motion for the possible electrons in the first row of the periodic elements. We will denote the second electron as $1e^{02}$. As noted before, we denote the point electrons (two in this case) of the first shell as $1e^{0i}$ where i have values 1 and 2. These two electrons with zero dimension manifolds are orbiting in a single orbit around the nucleus.

For the helium's atom, there are two electrons orbiting around the nucleus. The electrical field of the two electrons balances with the two positive charges of the nucleus. In this case, the element has two electrons, each with four points of spatial-connection, meeting the required eight points of spatial-connection for the electrons, creating a noble element. Helium establishes stability in the motion. There are only two electron manifolds possible for the first row, moving in a single orbit around the nucleus. The first shell is full (saturated) with two permissible electrons forming the first noble element helium (He) of the first row of the periodic table.

**Representation of the orbit of the first shell**

In the first shell, there are 2 electron manifolds acting as particles and orbiting around the nucleus. There is only one orbit for the first shell. The lines joining the point manifolds to the nucleus are perpendicular to the orbit. Thus, the orbit for the point manifolds in the first shell and also in the subsequent rows of elements is similar to a planetary orbit. The orbit of the two point manifolds is simple, and it adjusts to accommodate other multiple orbits of the higher number of points connected manifolds.

## 5.5.2 Atomic elements of the Second and Third Rows in Periodic Table

The second manifold of an electron is a curve (line). From Figure 5.1 it follows that there are six possible two points connected manifolds. These manifolds are of one dimensional curve (line) and are associated with electrons. For the second row, we represent the six curve manifolds as: (1, 2), (1, 3), (1, 4), (3, 4), (2, 4), and (2, 3). Let us call these manifolds as MC1, MC2, MC3, FC1, FC2, and FC3.

It is possible to have a different order of manifolds selection than the one presented above. This will also be the case for the manifolds in other rows. There is no reason to make the one order of selection of manifolds significant over the other, as the order depends on the geometry (and physics), and one is free to choose a different order of geometrical representation. We will make a selection of the order for a specific type of manifolds based on the empirical facts available from physics. In the above presentation, we have selected the order for the possible sets of manifolds, based on the occurrence of the electric and magnetic fields and directions of motion from the (two) types of manifolds. We will define the types of manifolds in Subsection 5.5.3.

The manifolds FC1, FC2, and FC3 include two spatial-points explicitly and do not have the electron's material point (1) implicitly. These manifolds are different in nature than those with explicit material point, as is the case with MC1 and others in which each of them include the electron's material point (1) and their motions become obvious.

The noted 6 manifolds orbit around the nucleus in the second shell. They demand to have the corresponding number of positively charged protons. These 6 electron manifolds are orbiting around the nucleus in 3 orbits under the electrical field and remain in the motion. The work done by the manifolds in their closed path of the orbit is zero. This follows from the Maxwell's Equation (5.2.2) as they are in the steady state motion and with no magnetic field in the region.

In the second shell, there are two types of manifolds. The first set of manifolds is with zero dimensions. The second set of manifolds is with one dimension curve (line). We will denote these two sets of electron manifolds, respectively, $2e^{0i}$ where i is variable and have values running from 1 to 2 that follows from the Subsection 5.5.1, and the curve manifolds as $2e^{1j}$ where j is variable and has values running from 1 to 6, that follows from the discussion of this Subsection. Also note that the number of electrons in the atomic orbits of the element are additive.

We will use the same reasoning in other Subsections and will not repeat it in the discussion of this Subsection in the others to follow. Due to the steady state motion, the negative electrical field is in the balance with the positive field of the nuclei and the elements are in neutral field conditions.

In the second row of elements, the first two point connected electron manifolds meet the eight point spatial-connections requirement for the orbit as discussed in Subsection 5.5.1. These two electron manifolds move in one orbit.

The same way, the second (and the third) row(s) of elements has three possible orbits. The curve manifolds will fill the three orbits. The first three manifolds MC1, MC2, and MC3 will occupy the three orbits. After that, the remaining curve manifolds FC1, FC2, and FC3 will start adding the second curve manifolds to meet the eight point spatial-connection requirements. Thus, the (MC1, FC1), (MC2, FC2), and (MC3, FC3) manifolds will fill the three orbits.

A selection of manifolds occupying the orbit is done, for example (MC1, FC1), so that both the manifolds together incorporate with possible curves that are made with all complementing points from the possibility of four points of connection associated with the electrons. Thus, in the orbit with manifold MC1 is with (1, 2) points, then the 2nd occupying manifold in the same orbit is FC1 has (3, 4) points. We will make the similar selections for the surface and volume manifolds in the following rows, and will not repeat how the manifolds are selected.

The last atomic element of the row, filled with six electron manifolds, gets fully saturated. Neon (Ne) is the last atomic element with fully saturated orbits. This reasoning applies to the other rows and we will not repeat it in the discussion to follow for the surface and volume manifolds.

There are a total of eight atomic elements in the second row of periodic table. The last element of the second row, neon (Ne), filled with all the possible number of manifolds in all possible openings of the possible orbits meet the eight point spatial-connections requirements. Neon turns into the noble element of the second row of the periodic table.

There are also manifolds with reverse orientation available with a second set of electrons, having the curves (2, 1), (3, 1), (4, 1), (4, 3), (4, 2), and (3, 2) creating a third shell and have a total of 8 elements in the third row of the periodic table. In the third row, we have a similar set of manifolds moving in the orbits around the nucleus. They are similar to the one in the second row, and represent the similar characteristics. So we will represent them as $3e^{0i}$, $3e^{1j}$ for the third shell. The third row also has eight elements. The last element of the row is another noble atomic element—argon (Ar).

The above noted atomic elements are for the first three rows of the periodic table. They cover all possible points and curves manifolds permissible with four points of spatial-connections for the electrons in the first three rows.

**Representation of the orbits of the second and third shells**

In the second and third shells, the electron manifolds have structures of particles and curves. The orbit related to the point manifolds is well defined in the Subsection 5.5.1 for the first shell. There are 3 more orbits for the curves and total of 4 orbits for the second and third shells.

The available geometrical points for the spatial-connection of electrons are adequate to have the motion without requiring any additional geometrical or temporal points of connection for the orbital motions. The lines joining the midpoint of three curves—(1, 2), (1, 3), and (1, 4)—to the nucleus are normal to the curve manifolds. These curve manifolds are orbiting along the tangent to these curves, and they are perpendicular to each other. Each orbit remains independent, adjusts to accommodate the other higher number of points connected manifolds, and continues to do so for the other rows.

## 5.5.3 Atomic elements of the Fourth and Fifth Rows in Periodic Table

The third manifold of an electron is a surface with three points of spatial-connections. The surface manifold in motion around the nucleus does not have the sufficient number of points required to make connections towards the nucleus and also in the direction of orbital motion which is along the tangent to its orbit. So the surface manifold needs to requisite additional points of connection. The required points of connections are available from the six points of spatial-connection for the electrons. The Figure 5.3 represents a geometrical structure of six points connected electrons, and will derive a possible number of manifolds for the fourth and fifth rows of the periodic table.

Before we find the information on the possible number of three points connected manifolds, let us comment first on the similarity with the fourth manifold of an electron, which is a volume with four points of spatial-connections that occurs in the sixth and seventh rows of the periodic table. These manifolds require an additional two points as the six points of connection is inadequate to describe their motions. The required additional two points are available from the eight points of spatial-connection of the electrons. We will discuss this case in the Subsection 5.5.4.

To add two points of connections for the motion of electron, let us extend the tetrahedron with four vertices to the octahedron with 6 vertices and 8 faces as shown in the Figure 5.3. Recall from the topology that the octahedron can transform from the six vertices to a four vertices tetrahedron and can continue to reduce its manifold characteristics to a surface, a curve, and a point without the loss of its connections from its geometrical structure.

Recall from physics that some of the elements of the fourth row have properties revealing magnetic fields that take place when the grains of atoms are oriented in specific directions and unpaired electrons are in motion. This physical property provides us the empirical fact that one of the vertices should represent time and other the space that associate with the motion of electrons. The space and time participate in the motion of electrons. But only the space (in the present discussion) unites with the material of electrons. The time remains absolute—Newtonian—and does not directly participate in the magnetic field. So, let us consider the vertices number 6 to denote the time, and 5 the spatial point. Let us look for a set of possible three points connected manifolds without the time, or its equivalent the vertices 6.

From Figure 5.3, it follows that there are two types of three points connected manifolds. First, there are 6 three points connected manifolds with one material (1) and the other two geometrical points are (1, 2, 3), (1, 2, 4), (1, 3, 4), (1, 3, 5), (1, 2, 5), and (1, 4, 5). Let us call them: MS1, MS2, MS3, MS4, MS5, and MS6. Second, there are 4 three points connected manifolds consisting of three spatial (geometrical) points, namely, (2, 3, 4), (2, 3, 5), (2, 4, 5), and (3, 4, 5). Let us call them: FS1, FS2, FS3 and FS4. There is no other three points connected manifold possible for the electron.

The electron manifolds FS1, FS2, FS3, and FS4 include three spatial points explicitly and represent field surfaces, and the physical material of electron appears implicitly. Their contributions in the orbital motion with other manifolds produce a number of unpaired electrons. These manifolds and the unpaired electrons introduce a magnetic field in the orbital motions of the electrons in the elements. From these observations, we define two types of manifolds as follows:

* A material manifold has two or more spatial points in its structure with one of them directly associates explicitly with a material point of the electron, and in the present study it pertains to point (1), representing a curve, or a surface, or a volume in the discussion.

* A field manifold has two or more spatial points of connection in its structure, and with no explicit material point from the electron, representing a curve, or a surface, or a volume in the discussion.

* We will callout, respectively, the material manifold with first letter M, and the field manifold with first letter F; second letter to show its type: C to denote a curve, S to denote a two dimensional surface, and V to denote a three dimensional volume; the third, and the fourth used for the sixth and seventh rows, expressed in numerals to identify with the list of the manifolds they represent.

Recall that the field manifolds implicitly include material and originate magnetic and electromagnetic fields—including radiations—for some of the atomic elements. The material manifolds explicitly include materials with associated fields and spatial points of connections.

Note that in the Figure 5.3, the curves (2, 4) and (3, 5) cross each other and appear in the manifold pairs: (MS2, MS4), (FS1, FS2), and (FS3, FS4). We need to arrange 10 manifolds so that the crossing curves do not appear on the same orbit, but will provide the needed eight points of spatial connections. Also an arrangement of these manifolds needs to be such that it creates unpaired electrons in the orbital motion in the corresponding atomic element that produces the magnetic field in it.

There are 10 three points connected surface manifolds. These manifolds are moving on 5 orbits around the nucleus. Based on the empirical facts of the elements, let us arrange them in two sets of 5 manifolds as (FS1, MS1, FS2, MS2, MS3) and (FS3, MS4, FS4, MS5, MS6).

One can arrange these manifolds in other order, such as, (MS1, FS1, MS2, FS2, MS3) and (MS4, FS3, MS5, FS4, MS6), which does not alter the final results of the discussion, as long as the manifolds do not cross as noted before. This note also applies to the other cases represented in the following subsections.

In this list, the first 5 manifolds starting from FS1 to MS3 occupy one by one the five orbits and then repeat the process to occupy the orbits with the second manifolds starting from FS3 to MS6 till all the five orbits are filled with 2 electron manifolds.

To distinguish these surface manifolds with top and bottom for the fourth and fifth rows of the periodic table, we can represent them with + and − surfaces, as + (1, 2, 4) and − (1, 2, 4), as done with positive and negative charges in the classical (Coulomb's) study of motion of electrons. We will not limit ourselves with the + and − representation, but will follow the one done by Maxwell. We have the following presentation possible to identify them clearly without attaching them with the positive and negative signs.

One can interpret the surface manifolds, however, to represent of having a circulation that must satisfy the Maxwell's Equation (5.2.3) for all points on the planetary orbital motion of electrons as they produce the magnetic field. Let us incorporate the noted physical property with geometrical property of the magnetic field lines to either counterclockwise or to clockwise presentation of the surface. To distinguish them, we will present these two sides of the surfaces with points (1, 2, 4) and (1, 4, 2) to represent the top and bottom of the same surface when observed from the top of its normal producing a counterclockwise connection of the curves (lines) enclosing the surface.

These 10 two dimensional manifolds of electrons will have their local top surface facing the nucleus in the fourth row. We will denote these electrons manifolds as $4e^{2k}$ where k is variable and has values 1 to 10. The fourth row of elements include $4e^{0i}$, $4e^{2k}$ and $4e^{1j}$, respectively, with zero dimensions—point manifolds, two dimensional surfaces—three points connected surface manifolds, and one dimensional curves—two points connected curve manifolds of the electrons, where i have values 1, 2; k have values 1 to 10 and j have values 1 to 6. Thus, the fourth shell has a total of 18 possible elements. The 18th element of the fourth shell gets fully saturated with all possible electrons and it is known as Kr (krypton) and its atomic number is 36.

For the fifth row of the periodic table, the 10 bottom based possible surfaces with one material point and two spatial points manifolds are (1, 3, 2), (1, 4, 2), (1, 3, 4), (1, 5, 3), (1, 5, 2), and (1, 5, 4). As before, let us call them as MS1, MS2, MS3, MS4, MS5, and MS6. Second, the three points connected manifolds consists of three spatial points are (2, 4, 3), (2, 5, 3), (2, 5, 4), and (3, 5, 4). Let us call them as FS1, FS2, FS3, and FS4.

As before, there are 10 three points connected surface manifolds for the fifth row. These manifolds are orbiting on 5 orbits around the nucleus. Based on the empirical facts of the elements, one of the arrangements for the manifolds is (FS1, MS1, FS2, MS2, MS3) and (FS3, MS4, FS4, MS5, MS6). There are a total of 18 atomic elements in the fifth row. As before, we will represent the electron manifolds in fifth row as $5e^{0i}$, $5e^{2k}$, and $5e^{1j}$. The last element of the fifth row is xenon (Xe) and its atomic number is 54.

### Representation of the orbits of the fourth and fifth shells

In the fourth and fifth shells, the electron manifolds have structures of particles, surfaces and curves. The Subsections 5.5.1 and 5.5.2 show the orbits related to the point and curve manifolds. We will present here only the case of the fourth shell with surface manifolds, as the fifth shell is a repetition of the fourth one except for the orientation, which does not affect the orbits representation.

There are 5 orbits in fourth shell for 10 surface-manifolds. These manifolds consist of 6 material manifolds, MS1, MS2, MS3, MS4, MS5, and MS6 and 4 spatial manifolds, FS1, FS2, FS3, and FS4. There are three sets of manifolds—(MS2, MS4), (FS1, FS2), and (FS3, FS4)—which have intersecting edges (2, 4) and (3, 5) and they are not permitted to be on the same orbits.

To establish the orbits, let us consider that the octahedron of Figure 5.3 represents the geometry of connections and also the physical characteristics of associated fields caused by the motion of electrons. Thus, the curves (1, 2) and (1, 3) denote the directions of the electric and magnetic fields, and (1, 4) the direction of motion of the electrons. Let us consider that the geometrical and physical properties of the octahedron establish the orbital connections with nucleus of the elements. The intersection point of the curves of manifolds denotes the curved geometry and physical manifolds, but does not create any physical interference associated with the fields in the motion of manifolds orbiting around the nucleus, as these physical manifolds are independent and are not constrained by the geometry of the octahedron. We will use the same geometrical and physical properties for the cube in the representation of the orbits in the sixth and seventh shells, and will not repeat the reasoning.

Based on the geometry of the octahedron of the Figure 5.3, surface manifolds FS2 and FS4, and their surfaces, (2, 3, 5) and (4, 3, 5) have a normal (2, 4) to their common curve (3, 5). Let us consider that the normal (2, 4) is facing towards the nucleus and intersects at the midpoint M of the curve (3, 5). Then, the curve (3, 5) is a tangent to the surfaces (2, 3, 5) and (4, 3, 5) and the surface manifolds. Thus, the manifolds FS2 and FS4 are orbiting around the nucleus. Their orbit passes through the point M, goes around the nucleus and then return to be tangent to the curve (3, 5) and closes at the point M. Likewise, the remaining two field manifolds FS1 and FS3 are orbiting along the second diagonal curve.

Similarly, the six material manifolds are orbiting along the three perpendicular orbits. These orbits are along the electric and magnetic fields and in the direction of motion of the electron.

For example, let us consider the manifolds MS1 and MS4 and their surfaces, (1, 2, 3) and (1, 3, 5), have a normal (3, 4) to the common curve (1, 3) representing an electric field. As before, let us consider that the normal (3, 4) is facing the nucleus. Then we can say that the curve (1, 3) is a tangent to the surfaces (1, 2, 3) and (1, 3, 5) and the two surface manifolds. In the motion of electrons, these two manifolds have an orbit along a tangent to the curve (1, 3) and are orbiting to enclose the nucleus. Likewise we can present the details for the remaining four material manifolds orbiting along (1, 2) and (1, 4). These three orbits are perpendicular to each other.

These are five orbits along which these surface manifolds are orbiting around the nucleus. Earlier we have arranged the manifolds so that each of them falls on a noted orbit and avoid interference with other orbits in their motions. Each orbit remains independent and accommodates the point and curve manifolds to have their motions in the fourth and fifth shells. There are respectively, 5 orbits for the surface, 3 for the curve, and 1 for the point manifolds, and a total of 9 orbits for the fourth and fifth shells.

We will keep the same order for the surface manifolds in the Subsection 5.5.4 for the sixth and seventh rows of the atomic elements.

## 5.5.4 Atomic elements of the Sixth and Seventh Rows in Periodic Table

The four-point connected volume manifold with six-point connected electron is not feasible to discuss the motion under the octahedron structure as the time will either appear in the manifold structure or in the spatial connection to the nucleus, or in the direction of motion. So, we will discuss the volume manifolds with eight points of spatial-connected electrons.

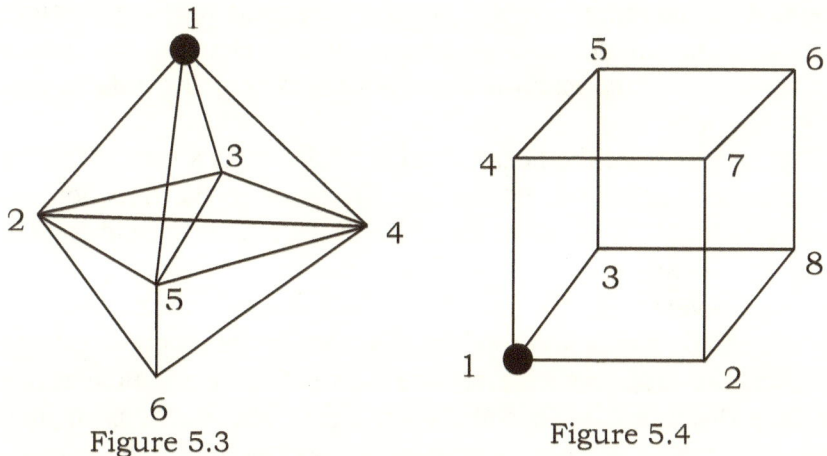

Figure 5.3          Figure 5.4

To add two points of connections to the electrons' motions, let us extend the octahedron with six vertices to hexahedron with 8 vertices and 6 faces as shown in the Figure 5.4. Note that topologically the eight vertices hexahedron can transfer to six vertices octahedron or four vertices tetrahedron, and so on.

Let us recall that the last three elements of the sixth and most of the elements of seventh rows are radioactive and some are short lived. They radiate with no external influence indicating that the atomic elements are functions of time and have electromagnetic radiative fields. Our interest is in the geometrical configuration of the elements and not in the radiation. So, we will focus on the geometrical structure of the elements. These physical properties, however, inform us the empirical facts that one of the vertices of the Figure 5.4 represents the time and the basic structure of the manifold includes electromagnetic fields associated with the space and matter. The spatial vertices participate to form a volume while the time participates in the motion and in the radiation. We will distinguish the time as a separate entity as it was the case for the fourth and fifth rows. So, we will use only 7 vertices, and discuss and derive the volume manifolds of electrons.

The hexahedron is a true platonic solid with 8 vertices, 12 edges of equal length, 6 square faces perpendicular to each other, and 3 faces that meet at each vertices, like electrical and magnetic fields and the direction of motion are. It is a symmetrical solid with eight points of connection for an electron, so selected atomic elements are functions of the number of the manifolds and not necessarily the order in which we select them.

Let us represent the hexahedron vertices (1) as a material point of the electron. As before, let the line (1, 2) be in the direction of an electric field, (1, 3) be in the direction of the magnetic field and (1, 4) be in the direction of motion. Thus, the three vertices 2, 3, and 4 directly associate with the material of the electron in the motion.

In a steady state motion of the electrons in the atomic elements of the sixth and seventh shell associate with the space and time as it was the case discussed in the Subsection 5.5.3. The electrons in motion unite with the neighboring spatial-points and the time. The vertices 5, 7, and 8 are geometrical points and associate with the motion of electrons. If the set of these three points (5, 7, 8) directly associate with point 1, then the volume manifold (1, 5, 7, 8) is similar, but not identical, to the manifold (1, 2, 3, 4).

From the motion of electrons, we will focus on the four points connected volume manifolds for the atomic elements. We will not consider the time vertex number 6 in the deduction of the volume manifolds.

From Figure 5.4 it follows that there are 14 local geometrical manifolds possible with four-points connected volumes. To maintain a clear picture of 14 volumes we prepare a list without drawing lines to any of the four vertices.

**A list of the 14 volumes is as follows:**

- Volume manifold with one material point and 3 associated (field) points: (1, 2, 3, 4), we will call this manifold as MV1.
- Volume manifolds with one material and 3 derived spatial-points: (1, 5, 7, 8), we will call this manifold as MV2.
  Based on the union of space and matter with the three spatial-points are equivalent to the volume of four field points. So, MV2 will be on the same orbit as that of MV1. These are 2 manifolds.

- Volume manifolds with one material (1), one spatial, and two associated points are: (1, 2, 3, 5), (1, 3, 4, 7), (1, 4, 2, 8), (1, 2, 3, 7), (1, 3, 4, 8), and (1, 4, 2, 5). These are 6 manifolds.
  List these manifolds as: MV11, MV12, MV13, MV14, MV15, and MV16.
- Volume manifolds with two associated and two spatial points are: (2, 3, 8, 5), (3, 4, 8, 5), (2, 4, 7, 5), (2, 3, 8, 7), (3, 4, 8, 7), and (2, 4, 7, 8). These are 6 manifolds.
  List these manifolds as: FV21, FV22, FV23, FV24, FV25, and FV26.

There are other 9 volumes (manifolds) with a material point (1), and associated point, and two spatial points. These volumes are: (1, 2, 5, 8), (1, 3, 5, 8), (1, 4, 5, 8), (1, 2, 5, 7), (1, 3, 5, 7), (1, 4, 5, 7), (1, 2, 7, 8), (1, 3, 7, 8), and (1, 4, 7, 8). These are deceiving volumes. Why these manifolds are deceiving is discussed below.

The above noted volumes are geometrically possible, but are ambiguous and not well defined manifolds; they include overlapping associated and spatial curves from the same field or points from the same domain of construction of the hexahedron making them an indecisive manifold to consider in the study. For example, the volume (1, 2, 5, 8) includes two surfaces (1, 2, 5) and (1, 8, 5) having curves (1, 2) and (1, 8) with vertices 2 and 8 falling on the same electrical field surface, and the curves (1, 5) and (8, 5) with vertices falling on the plane of time with the manifold having indecisive physical structure. Similarly the volume (1, 2, 5, 7) includes an electromagnetic surface (2, 5, 7) with two curves (2, 5) and (2, 7) with vertices falling on the plane of time, and so on. Thus, these are volumes, but not well-defined manifolds. They are ambiguous and confusing to define its orientation towards the nucleus. They are unacceptable volumes to consider them as manifolds.

There are also other possible combinations of four vertices volumes (manifolds), but they either represent the same volume manifolds included in the above list, or they do not create a closed volume with the associated surfaces, or do not have a feasible volume from the available surfaces.

To have a clear understanding of these four-point (vertices) connected 14 volumes, we note of their geometrical characteristics. These 14 volume manifolds enclose the neighboring space. Each of the volume consists of 4 vertices, 6 edges, and 4 surfaces. Each of the surfaces is a combination of three edges. Depending on the volume under discussion; there are two or three external, and one or two internal surfaces, respectively, from the hexahedron. These volume manifolds split the hexahedron, and at least one of the internal and one of the external surfaces are directed towards the nucleus in their orbital motions.

Let us recall that the electric, magnetic fields, and the direction of motion of the electron are perpendicular. The four points connected manifolds of the electrons are under a steady state motion, satisfy the Maxwell's Equations (5.2.3) and (5.2.4) as the electrons are moving under no currents, no magnetic field, and are not directly dependent on time.

For the sixth row of the periodic table of elements, the volumes are from the inside of the hexahedron facing the nucleus as it provides stable elements compared to the one from the external surface having radiation, which is for the seventh row of the periodic table. Note that the external surfaces depend on the time, while the inside diagonal surfaces depend only on the electric and magnetic fields created by the motion of electrons.

Note that these volume manifolds include three sets of crossing curves: [(1, 5) and (7, 8)], [(1, 7) and (5, 8)], and [(1, 8) and (5, 7)]. These curves, however, are not in the same planes (surfaces), as the one appeared in the fourth and fifth rows which were located in the same plane. As we noted for the fourth row of elements that two crossing curve manifolds produce the magnetic field, these three sets of crossing curve manifolds produce the elements with electromagnetic fields that radiate as time participates in separation of these curves. These distant curves indicate that the material and spatial manifolds are orbiting at different distances from the nucleus, like the planets are orbiting at different distances from the sun. Also, as the time appears in the crossing manifolds, they create radiation for the most (except 3) of the elements in the seventh row.

These manifolds are with the inside and outside surfaces for the sixth and seven rows, respectively. We will denote these electron manifolds as $6e^{3m}$, and $7e^{3m}$, where m is variable and has values of 1 to 14. These 14 volume manifolds—MV1, MV11, MV12, MV13, FV21, FV22, and FV23—will occupy one by one the 7 orbits. Recall that in the seventh row of the elements, all elements release electromagnetic radiation except for the element numbers 90, 91, and 92. Our order of manifolds selection is made such that that it meets these requirements for the non–radiating (3) elements in the seventh row.

Once all the seven orbits are occupied by a volume manifold, the remaining elements will repeat the process of occupying the orbits with the second set of manifolds in the available opening in the orbits. This is done by the manifolds MV2, MV14, MV15, MV16, FV24, FV25, and FV26. Thus, each of the 7 orbits is filled with 2 volume manifolds in the sixth and seventh rows.

The sixth shell includes $6e^{0i}$, $6e^{3m}$, $6e^{2k}$ and $6e^{1j}$ electron manifolds, respectively, with zero dimensions-points; three dimensional four points connected volumes; two dimensional three points connected surfaces; and one dimensional two points connected curves; where i have values 1, 2; m have values 1 to 14; k have values 1 to 10 and j have values 1 to 6. Thus, the sixth shell has a total of 32 possible atomic elements. The 32nd atomic element of the sixth row gets fully saturated with all possible electrons and it is known as radon (Rn) and its atomic number is 86.

For the seventh row of the periodic table, there are 14 external volume connected manifolds listed above. Again, the seventh row has similar 32 atomic elements. But, for this case, one of the exterior surfaces of the volume manifolds is facing the nucleus. Some of the electron manifolds are time depend and are in highly unstable conditions producing a radioactive and short lived elements. We will not go into radioactive properties in this book, but will represent the electron manifolds of the row as $7e^{0i}$, $7e^{3m}$, $7e^{2k}$ and $7e^{1j}$; the last element of the row is the noble element No 118, which was discovered in 2015 and IUPAC (International Union of Pure and Applied Chemistry) has accepted as a noble element. In the attached list, there are 4 new elements IUPAC has accepted their existence and in the list of periodic table their proposed names and symbols are included, which need to be approved by IUPAC members.

In the sixth and seventh shells, the electron manifolds have structures of particles, curves, surfaces and volumes. The Subsections 5.5.1, 5.5.2, and 5.5.3 show the orbits related to the point, curve and surface manifolds. We will discuss only the case for the sixth shell.

Note that the seventh shell is highly unstable, have some repetitions of the sixth shell, and depend on time intensively. The seventh shell, however, has the same number of orbits as sixth shell, except the orientation that does not affect the representation of the orbits, so we will not repeat in the following.

**Representation of the orbits of the sixth and seventh shells**

There are seven orbits in the sixth shell for 14 volume manifolds. These manifolds consist of 8 material manifolds: MV1, MV2; MV11, MV12, MV13, MV14, MV15, and MV16, and 6 spatial manifolds: FV21, FV22, FV23, FV24, FV25, and FV26.

The volume manifolds MV1 and MV2 are with a material point and three associated points or three derived points. These manifolds are orbiting on the same orbit, and are similar to a point manifold. To avoid any interaction between the point and volume manifolds, and to define a normal to a point and a volume manifolds is hard, we will consider that the MV1 and MV2 are orbiting in a plane perpendicular to the plane of the point manifolds orbit.

The volume manifolds are derived from the cube of Figure 5.4, with vertices 1 and 6 representing material and time, respectively. These manifolds are orbiting around the nucleus of an element. To accommodate the nucleus of the element around which the manifolds are orbiting, and to establish the orbits of the manifolds, let us consider, as before, that the cube of Figure 5.4 represents geometrical connections and also the physical characteristics of the associated fields caused by the motion of electrons. The geometry of the cube permits to consider the vertices 1 and 6 as geometrical points. Let us consider the geometry and physical properties associated with the cube to permit to establish a connection with the nucleus of the element around which the electron manifolds are orbiting. We will discuss the orbits in two parts.

First, let us discuss one of the three orbits for the six material manifolds. The manifolds MV11 and MV14, and their geometrical volumes (1, 2, 3, 5) and (1, 2, 3, 7) have their interior surfaces (2, 3, 5) and (2, 3, 7) facing the nucleus. A normal to these two surfaces is facing towards the center of the nucleus. Let us denote this normal as line N1. Line N1 is perpendicular to the surfaces (2, 3, 5) and (2, 3, 7) and so to the volumes (1, 2, 3, 5) and (1, 2, 3, 7), and manifolds MV11 and MV14. These two surfaces create a diagonal plane (2, 3, 5, 7) in the cube. Let us denote this plane as P. N1 is perpendicular to plane P. Then, plane P includes a part of the orbit of the two manifolds orbiting around the nucleus. To define a complete orbit to these manifolds, let us select a line, T1, in the plane P and perpendicular to the lines (2, 3) and (5, 7) meeting at midpoints M1 and L1, respectively. By geometrical construction, line T1 is perpendicular to line N1, and is along the tangent to the manifolds MV11 and MV14 and denotes part of the orbit to these two manifolds. (In the second part, we will use a second choice in the plane P for an orbit to other spatial manifolds.) Thus, manifolds MV11 and MV14 are orbiting around the nucleus. The orbit passes through the point M1 on the top of the cube, goes around the nucleus and return at the bottom of the cube at L1, be a tangent to the line T1 and closes at the midpoint M1.

Similarly, the manifolds MV12 and MV15 and their geometrical volumes (1, 3, 4, 7) and (1, 3, 4, 8) are orbiting around the nucleus. The orbit passes through the midpoint of the curve (4, 3), say point Q1, goes around the nucleus and return to the midpoint of the curve (7, 8), be a tangent to the plane (3, 4, 7, 8), and closes at the midpoint Q1.

Likewise, the manifolds MV13 and MV16 and their geometrical volumes (1, 4, 2, 8) and (1, 4, 2, 5) are orbiting around the nucleus. The orbit passes through the midpoint of the curve (2, 4), say point R1, goes around the nucleus and returns to the midpoint of the curve (8, 5), be a tangent to the plane (2, 4, 5, 8), and closes at the midpoint R1.

Second, we will find other three orbits for the remaining 6 spatial manifolds. Let us recall that these manifolds include three sets of crossing curves: [(1, 5) and (7, 8)], [(1, 7) and (5, 8)], and [(1, 8) and

(5, 7)]. These curves, however, are not in the same plane (surface), as the one appeared in the fourth and fifth rows which were located in the same plane. These three sets of curves separations are caused by the implicit time (6) in the curves, and no explicit material (1) in the surface manifolds. The time and material appear implicitly in the manifolds, causing the curves to separate.

The crossing curves (2, 3) and (1, 8) are in the diagonal plane (1, 8, 6, 4), require them having unacceptable material vertex (1) and the time vertex (6) to participate in the orbit. But, there cannot be any orbit along this diagonal plane of the cube as we are not using the time vertex 6. The same reasoning applies to other two diagonal planes with material (1) and time (6) vertices.

Thus, the orbital motion forces the material curves (1, 5), (1, 7), and (1, 8) to be replaced with the curves parallel to the set of curves—(7, 8), (5, 8) and (5, 7) as crossing curves. For example, the curve (2, 3) replaces the curve (1, 8), which is a curve parallel to the curve (5, 7) in the top plane and creates a diagonal plane (2, 3, 5, 7) for the orbiting manifolds. This is the same plane P we used to discuss in the First case of the material manifolds. The same reasoning applies to the other two diagonal planes.

Plane P also includes part of the orbit of the spatial manifolds FV21, FV24 and the volumes (2, 3, 8, 7) and (2, 3, 8, 5). To establish the orbit, let us recall that in the First case we selected the orbit to be perpendicular to the lines (2, 3) and (5, 7). So, in this case, to avoid any interference with the material orbit, we select the orbit to be parallel to (2, 3) and (5, 7), or its equivalent perpendicular to the lines (2, 7) and (3, 5). Let us apply similar reasoning to the volumes (2, 3, 8, 5) and (2, 3, 8, 7) having interior surfaces (2, 3, 5) and (2, 3, 7) facing the nucleus. Note that the surfaces (2, 3, 5) and (2, 3, 7) appear to be the same as the one noted in the First case, but in reality these surfaces are different from the different manifolds and have different physical characteristics then the one noted before. A normal to these two surfaces is along the center of the nucleus and let us denote as line N2. The line N2 is perpendicular to the surfaces (2, 3, 5) and (2, 3, 7), to the volumes (2, 3, 8, 5) and (2, 3, 8, 7), manifolds FS21 and FS24, and to the plane P. As noted before, P includes part of the orbit of the manifolds. To define a complete orbit, let us select a line, T2, perpendicular to the line (2, 7) and (3, 5) meeting at midpoints M2 and L2, respectively. By geometrical construction, line T2 is perpendicular to line N2, is along the tangent to the manifolds FS21 and FS24, and denotes part of the orbit to these manifolds. Thus the manifolds FS21 and FS24 are orbiting around the nucleus. The orbit passes through point M2, goes around the nucleus to return at the midpoint L2 of the edge (3, 5), be a tangent to line T2, and closes at the midpoint M2.

Similarly, the manifolds FV22 and FV25 and their geometrical volumes (3, 4, 8, 5) and (3, 4, 8, 7) are orbiting around the nucleus. The orbit passes through the midpoint of the curve (4, 7), say Q2, goes around the nucleus and returns to the midpoint of the curve (3, 8), be a tangent to the plane (3, 4, 7, 8), and closes at midpoint Q2.

Likewise, the manifolds FV23 and FV26 and their geometrical volumes (2, 4, 7, 5) and (2, 4, 7, 8) are orbiting around the nucleus. The orbit passes through the midpoint of the curve (4, 5), say R2, goes around the nucleus and returns to the midpoint of the curve (2, 8), be a tangent to the plane (2, 4, 5, 8), and closes at midpoint R2.

There are a total of 16 orbits in the sixth and seventh shells. These orbits are: 7 for the volume manifolds, 5 for the surface manifolds, 3 for the curve manifolds, and 1 for the point manifolds.

Note that there is no geometrical restriction on the size of the cube and their edges. So, the geometrical and physical properties of a cube permit easily to demonstrate the electron manifolds orbiting around the nucleus.

## 5.5.5  Comments on atomic elements in periodic table

There are four sets of geometrical configuration of manifolds in seven rows of atomic elements in the periodic table. These manifolds depend only on the spatial-connections of the electrons and Maxwell's equations. The manifolds and their orbits are independent of the quantum numbers—atomic number, azimuthal angle, orbital number, magnetic moment; or the (Schrödinger's) wave equations—related to electrons motions in atoms without having any clear physical or mathematical properties describing their association with motion of electrons in atomic orbits. Due to these noted limitations, we have derived the structure of atomic elements in terms of number of manifolds of electrons orbiting around the nucleus that simply follows the shape of their geometrical configuration.

The representation of the orbits neither needs to follow the complex probability-density distribution nor to be intricate as is the case in quantum mechanics. The orbits are simple and easy to establish by using simple geometrical structures of tetrahedron, octahedron, and cube.

The periodic table presented in the following section has a well-defined order, constancy in the orbital motions and geometrical configurations of the atomic elements. Due to these facts, the periodic table presented here has a variation in their orders for two atomic elements in the sixth and seventh rows from the one constructed based on the variables of quantum mechanics.

As noted before, atomic elements Nos. 57 and 89 are volume manifolds while the atomic elements Nos. 71 and 103 are surface manifolds, shifting the order of the elements by one place from the periodic table currently in use. Only the experiment can decide the geometrical facts of the electrons orbiting around the nucleus, but theoretically their atomic weight and other physical and chemical characteristics remain the same.

The presentation delivered in this chapter to the world of chemists, physicists, and scientists definitely will hold the opposing views to the one how they have approached and developed the atomic elements. Based on their representations of elements, it is enticing to look for unstable, radioactive elements occurring in Actinide Series. They have developed the elements, going through complicated process, like americium, curium, berkelium, californium, and so on, as noted in Section 5.6 below.

At this stage, I do not blame their opposition to my presentation or I do not underestimate their contributions to develop the atomic elements of periodic table. To discover a new element is good, but based on a simple geometrical representation of the electrons is better—without addressing to the Mendeleev's or Moseley's methods, or without addressing to emitting light spectrum, or without any theory of quantum mechanics. It is worthwhile to visualize how the geometrical structured electrons are located in the elements.

## 5.6 Periodic table of atomic elements

One can express the periodic table of atomic elements in a matrix of 8 columns and 7 rows. It is, however, inconvenient to arrange all the elements in this format on an 6"× 9" page. So, we simply list all of 118 atomic elements with their name, number of electrons, and their configuration in terms of the associated geometrical manifold structure in the following table.

| Row Number | Name of atom | Number of Electrons in atomic element | Electron configuration in manifold structure | Comments |
|---|---|---|---|---|
| 1 | Hydrogen (H) | 1 | $1e^{01}$ | |
| 1 | Helium (He) | 2 | $1e^{02}$ | Noble gas (He) with 2 electrons |
| 2 | Lithium (Li) | 3 | $(He)\ 2e^{01}$ | |
| 2 | Beryllium (Be) | 4 | $(He)\ 2e^{02}$ | |
| 2 | Boron (B) | 5 | $(He)\ 2e^{02}\ 2e^{11}$ | |
| 2 | Carbon (C) | 6 | $(He)\ 2e^{02}\ 2e^{12}$ | |
| 2 | Nitrogen (N) | 7 | $(He)\ 2e^{02}\ 2e^{13}$ | |
| 2 | Oxygen (O) | 8 | $(He)\ 2e^{02}\ 2e^{14}$ | |
| 2 | Fluorine (F) | 9 | $(He)\ 2e^{02}\ 2e^{15}$ | |
| 2 | Neon | 10 | $(He)\ 2e^{02}\ 2e^{16}$ | Noble gas (Ne) with 10 electrons |
| 3 | Sodium (Na) | 11 | $(Ne)\ 3e^{01}$ | |
| 3 | Magnesium (Mg) | 12 | $(Ne)\ 3e^{02}$ | |
| 3 | Aluminum (Al) | 13 | $(Ne)\ 3e^{02}\ 3e^{11}$ | |
| 3 | Silicon (Si) | 14 | $(Ne)\ 3e^{02}\ 3e^{12}$ | |
| 3 | Phosphorus (P) | 15 | $(Ne)\ 3e^{02}\ 3e^{13}$ | |
| 3 | Sulfur (S) | 16 | $(Ne)\ 3e^{02}\ 3e^{14}$ | |
| 3 | Chlorine (Cl) | 17 | $(Ne)\ 3e^{02}\ 3e^{15}$ | |
| 3 | Argon (Ar) | 18 | $(Ne)\ 3e^{02}\ 3e^{16}$ | Noble gas (Ar) |

| | | | | with 18 electrons |
|---|---|---|---|---|
| 4 | Potassium (K) | 19 | $(Ar)\,4e^{01}$ | |
| 4 | Calcium (Ca) | 20 | $(Ar)\,4e^{02}$ | |
| 4 | Scandium (Sc) | 21 | $(Ar)\,4e^{02}\,4e^{21}$ | |
| 4 | Titanium (Ti) | 22 | $(Ar)\,4e^{02}\,4e^{22}$ | |
| 4 | Vanadium (V) | 23 | $(Ar)\,4e^{02}\,4e^{23}$ | |
| 4 | Chromium (Cr) | 24 | $(Ar)\,4e^{02}\,4e^{24}$ | |
| 4 | Manganese (Mn) | 25 | $(Ar)\,4e^{02}\,4e^{25}$ | |
| 4 | Iron (Fe) | 26 | $(Ar)\,4e^{02}\,4e^{26}$ | |
| 4 | Cobalt (Co) | 27 | $(Ar)\,4e^{02}\,4e^{27}$ | |
| 4 | Nickel (Ni) | 28 | $(Ar)\,4e^{02}\,4e^{28}$ | |
| 4 | Copper (Cu) | 29 | $(Ar)\,4e^{02}\,4e^{29}$ | |
| 4 | Zinc (Zn) | 30 | $(Ar)\,4e^{02}\,4e^{210}$ | |
| 4 | Gallium (Ga) | 31 | $(Ar)\,4e^{02}\,4e^{210}\,4e^{11}$ | |
| 4 | Germanium (Ge) | 32 | $(Ar)\,4e^{02}\,4e^{210}\,4e^{12}$ | |
| 4 | Arsenic (Ar) | 33 | $(Ar)\,4e^{02}\,4e^{210}\,4e^{13}$ | |
| 4 | Selenium (Se) | 34 | $(Ar)\,4e^{02}\,4e^{210}\,4e^{14}$ | |
| 4 | Bromine (Br) | 35 | $(Ar)\,4e^{02}\,4e^{210}\,4e^{15}$ | |
| 4 | Krypton (Kr) | 36 | $(Ar)\,4e^{02}\,4e^{210}\,4e^{16}$ | Noble gas (Kr) with 36 electrons |
| 5 | Rubidium (Rb) | 37 | $(Kr)\,5e^{01}$ | |
| 5 | Strontium (Sr) | 38 | $(Kr)\,5e^{02}$ | |
| 5 | Yttrium (Y) | 39 | $(Kr)\,5e^{02}\,5e^{21}$ | |
| 5 | Zirconium (Zr) | 40 | $(Kr)\,5e^{02}\,5e^{22}$ | |
| 5 | Niobium (Nb) | 41 | $(Kr)\,5e^{02}\,5e^{23}$ | |
| 5 | Molybdenum (Mo) | 42 | $(Kr)\,5e^{02}\,5e^{24}$ | |
| 5 | Technetium (Tc) | 43 | $(Kr)\,5e^{02}\,5e^{25}$ | |
| 5 | Ruthenium (Ru) | 44 | $(Kr)\,5e^{02}\,5e^{26}$ | |
| 5 | Rhodium (Rh) | 45 | $(Kr)\,5e^{02}\,5e^{27}$ | |
| 5 | Palladium (Pd) | 46 | $(Kr)\,5e^{02}\,5e^{28}$ | |
| 5 | Silver (Ag) | 47 | $(Kr)\,5e^{02}\,5e^{29}$ | |
| 5 | Cadmium (Cd) | 48 | $(Kr)\,5e^{02}\,5e^{210}$ | |
| 5 | Indium (In) | 49 | $(Kr)\,5e^{02}\,5e^{210}\,5e^{11}$ | |

| 5 | Tin (Sn) | 50 | $(Kr)\, 5e^{02}\, 5e^{210}$ $5e^{12}$ | |
| 5 | Antimony (Sb) | 51 | $(Kr)\, 5e^{02}\, 5e^{210}$ $5e^{13}$ | |
| 5 | Tellurium (Te) | 52 | $(Kr)\, 5e^{02}\, 5e^{210}$ $5e^{14}$ | |
| 5 | Iodine (I) | 53 | $(Kr)\, 5e^{02}\, 5e^{210}$ $5e^{15}$ | |
| 5 | Xenon (Xe) | 54 | $(Kr)\, 5e^{02}\, 5e^{210}$ $5e^{16}$ | Noble gas (Xe) with 54 electrons |
| 6 | Cesium (Cs) | 55 | $(Xe)\, 6e^{01}$ | |
| 6 | Barium (Ba) | 56 | $(Xe)\, 6e^{02}$ | |
| 6 | Lanthanum (La) | 57 | $(Xe)\, 6e^{02}\, 6e^{31}$ | |
| 6 | Cerium (Ce) | 58 | $(Xe)\, 6e^{02}\, 6e^{32}$ | |
| 6 | Praseodymium (Pr) | 59 | $(Xe)\, 6e^{02}\, 6e^{33}$ | |
| 6 | Neodymium (Nd) | 60 | $(Xe)\, 6e^{02}\, 6e^{34}$ | |
| 6 | Promethium (Pm) | 61 | $(Xe)\, 6e^{02}\, 6e^{35}$ | |
| 6 | Samarium (Sm) | 62 | $(Xe)\, 6e^{02}\, 6e^{36}$ | |
| 6 | Europium (Eu) | 63 | $(Xe)\, 6e^{02}\, 6e^{37}$ | |
| 6 | Gadolinium (Gd) | 64 | $(Xe)\, 6e^{02}\, 6e^{38}$ | |
| 6 | Terbium (Tb) | 65 | $(Xe)\, 6e^{02}\, 6e^{39}$ | |
| 6 | Dysprosium (Dy) | 66 | $(Xe)\, 6e^{02}\, 6e^{310}$ | |
| 6 | Holmium (Ho) | 67 | $(Xe)\, 6e^{02}\, 6e^{311}$ | |
| 6 | Erbium (Er) | 68 | $(Xe)\, 6e^{02}\, 6e^{312}$ | |
| 6 | Thulium (Tm) | 69 | $(Xe)\, 6e^{02}\, 6e^{313}$ | |
| 6 | Ytterbium (Yb) | 70 | $(Xe)\, 6e^{02}\, 6e^{314}$ | |
| 6 | Lutetium (Lu) | 71 | $(Xe)\, 6e^{02}\, 6e^{314}$ $6e^{21}$ | |
| 6 | Hafnium (Hf) | 72 | $(Xe)\, 6e^{02}\, 6e^{314}$ $6e^{22}$ | |
| 6 | Tantalum (Ta) | 73 | $(Xe)\, 6e^{02}\, 6e^{314}$ $6e^{23}$ | |
| 6 | Tungsten (W) | 74 | $(Xe)\, 6e^{02}\, 6e^{314}$ $6e^{24}$ | |
| 6 | Rhenium (Re) | 75 | $(Xe)\, 6e^{02}\, 6e^{314}$ $6e^{25}$ | |
| 6 | Osmium (Os) | 76 | $(Xe)\, 6e^{02}\, 6e^{314}$ $6e^{26}$ | |
| 6 | Iridium (Ir) | 77 | $(Xe)\, 6e^{02}\, 6e^{314}$ | |

| | | | | |
|---|---|---|---|---|
| | | | $6e^{27}$ | |
| 6 | Platinum (Pt) | 78 | (Xe) $6e^{02}\,6e^{314}$ $6e^{28}$ | |
| 6 | Gold (Au) | 79 | (Xe) $6e^{02}\,6e^{314}$ $6e^{29}$ | |
| 6 | Mercury (Hg) | 80 | (Xe) $6e^{02}\,6e^{314}$ $6e^{210}$ | |
| 6 | Thallium (Ti) | 81 | (Xe) $6e^{02}\,6e^{314}$ $6e^{210}\,5e^{11}$ | |
| 6 | Lead (Pb) | 82 | (Xe) $6e^{02}\,6e^{314}$ $6e^{210}\,6e^{12}$ | |
| 6 | Bismuth (Bi) | 83 | (Xe) $6e^{02}\,6e^{314}$ $6e^{210}\,6e^{13}$ | |
| 6 | Polonium (Po) | 84 | (Xe) $6e^{02}\,6e^{314}$ $6e^{210}\,6e^{14}$ | |
| 6 | Astatine (At) | 85 | (Xe) $6e^{02}\,6e^{314}$ $6e^{210}\,6e^{15}$ | |
| 6 | Radon (Rn) | 86 | (Xe) $6e^{02}\,6e^{314}$ $6e^{210}\,6e^{16}$ | Noble gas (Rn) with 86 electrons |
| 7 | Francium (Fr) | 87 | (Rn) $7e^{01}$ | |
| 7 | Radium (Ra) | 88 | (Rn) $7e^{02}$ | |
| 7 | Actinium (Ac) | 89 | (Rn) $7e^{02}\,7e^{31}$ | |
| 7 | Thorium (Th) | 90 | (Rn) $7e^{02}\,7e^{32}$ | |
| 7 | Protactinium (Pa) | 91 | (Rn) $7e^{02}\,7e^{33}$ | |
| 7 | Uranium (U) | 92 | (Rn) $7e^{02}\,7e^{34}$ | |
| 7 | Neptunium (Np) | 93 | (Rn) $7e^{02}\,7e^{35}$ | |
| 7 | Plutonium (Pu) | 94 | (Rn) $7e^{02}\,7e^{36}$ | |
| 7 | Americium (Am) | 95 | (Rn) $7e^{02}\,7e^{37}$ | |
| 7 | Curium (Cm) | 96 | (Rn) $7e^{02}\,7e^{38}$ | |
| 7 | Berkelium (Bk) | 97 | (Rn) $7e^{02}\,7e^{39}$ | |
| 7 | Californium (Cf) | 98 | (Rn) $7e^{02}\,7e^{310}$ | |
| 7 | Einsteinium (Es) | 99 | (Rn) $7e^{02}\,7e^{311}$ | |
| 7 | Fermium (Fm) | 100 | (Rn) $7e^{02}\,7e^{312}$ | |
| 7 | Mendelevium (Md) | 101 | (Rn) $7e^{02}\,7e^{313}$ | |
| 7 | Nobelium (No) | 102 | (Rn) $7e^{02}\,7e^{314}$ | |
| 7 | Lawrencium (Lr) | 103 | (Rn) $7e^{02}\,7e^{314}$ $7e^{21}$ | |
| 7 | Rutherfordium (Rf) | 104 | (Rn) $7e^{02}\,7e^{314}$ $7e^{22}$ | |

| 7 | Dubnium (Db) | 105 | $(Rn)\ 7e^{02}\ 7e^{314}$ $7e^{23}$ | |
|---|---|---|---|---|
| 7 | Seaborgium (Sg) | 106 | $(Rn)\ 7e^{02}\ 7e^{314}$ $7e^{24}$ | |
| 7 | Bhorium (Bh) | 107 | $(Rn)\ 7e^{02}\ 7e^{314}$ $7e^{25}$ | |
| 7 | Hassium (Hs) | 108 | $(Rn)\ 7e^{02}\ 7e^{314}$ $7e^{26}$ | |
| 7 | Meitnerium (Mt) | 109 | $(Rn)\ 7e^{02}\ 7e^{314}$ $7e^{27}$ | |
| 7 | Darmstadtium (Ds) | 110 | $(Rn)\ 7e^{02}\ 7e^{314}$ $7e^{28}$ | |
| 7 | Roentgenium (Rg) | 111 | $(Rn)\ 7e^{02}\ 7e^{314}$ $7e^{29}$ | |
| 7 | Copernicium (Cn) | 112 | $(Rn)\ 7e^{02}\ 7e^{314}$ $7e^{210}$ | |
| 7 | Recently discovered and accepted by *IUPAC— Named: Nihonium (Nh) | 113 | $(Rn)\ 7e^{02}\ 7e^{314}$ $7e^{210}7e^{11}$ | |
| 7 | Flerovium (Fl) | 114 | $(Rn)\ 7e^{02}\ 7e^{314}$ $7e^{210}7e^{12}$ | |
| 7 | Recently discovered and accepted by *IUPAC— Named: Moscovium (Mc) | 115 | $(Rn)\ 7e^{02}\ 7e^{314}$ $7e^{210}7e^{13}$ | |
| 7 | Livermorium (Lv) | 116 | $(Rn)\ 7e^{02}\ 7e^{314}$ $7e^{210}7e^{14}$ | |
| 7 | Recently discovered and accepted by *IUPAC— Named: Tennessine (Ts) | 117 | $(Rn)\ 7e^{02}\ 7e^{314}$ $7e^{210}7e^{15}$ | |
| 7 | Recently discovered and accepted by | 118 | $(Rn)\ 7e^{02}\ 7e^{314}$ $7e^{210}7e^{16}$ | Recently discovered Noble gas with 118 |

| | *IUPAC—<br>Named:<br>Ogannesson (Og) | | | electrons |
|---|---|---|---|---|
| | * IUPAC = International Union of Pure and Applied Chemistry, who need to confirm the name and symbol | | | |
| | This cell is intentionally left blank | | | |

## 5.7   Conclusion

1.  In electrodynamics, there are 8 Maxwell's equations that permit to have 8 well defined points of connections for the motions of electrons. These points in motions of electrons associate with the neighboring space, its matter and fields creating manifolds—geometrical structures—starting from a point, to a curve, to a surface and to a volume.

2.  Following Moseley's findings, all the elements of the periodic table are arranged based on the number of protons in the elements (and not on the atomic weight as proposed by Mendeleev). The number of protons and electrons in the elements are the same.

3.  Electrons have a minimum of four points of connection causing to have spin in it. Electrons have a half spin. Due to its half spin, an electron's front observable side turns into non-observable back (obverse) side when it completes the full orbit and non-observable back (obverse) side turns into the front observable side after the second revolution.

4.  Electrons' spin, and their possible particle, curve, surface, and volume manifolds originate 7 rows (periods). And the electrons' four point connection in motion originate 8 columns (groups) in the periodic table.

5.  Only the Maxwell's equations and the connections of electrons are adequate to develop and present the unambiguous geometrical configuration of all the 118 atomic elements of periodic table. Out of these 118 possible elements, the recently discovered (artificially lab created) 4 elements are accepted by

IUPAC (International Union of Pure and Applied Chemistry) at the writing of this book.

6. Based on the above list of atomic elements including the recently developed and accepted elements by IUPAC, there are only 118 elements in the periodic table, based on the simplified geometrical manifold structure of the electrons.

7. There is a limit on the geometrical structure of connections of electrons and only 118 atomic elements are possible and no more.

## Epilogue

The tenet of a point particle has now departed. A particle with two or more geometrical points replaced it.

## Why?

**Because an electron at rest has two or more geometrical points of connection and, in motion it can have up to eight points of connection.**

For centuries, philosophers, mathematicians, physicists, and chemists have been, respectively, intrigued by the concepts associated with atomic elements and raised fundamental questions. Philosophically: How small a material can split to form an atom, a non-divisible material particle? Mathematically: Which mathematical characteristics are associated with the number of electrons in atomic elements? Physically: What are the physical structures of motions of electrons in atomic elements? And Chemically: Why are there similarities in chemical properties of atomic elements in the different rows of the periodic table?

To answer these questions let us recall that a physical theory is always provisional, in the sense that it is a model based on some hypothesis, basic concepts which cannot be proved. For many times the experimental results would agree with the theory, but one cannot be sure of that the next time for a new experiment that will not contradict the theory; or there are experiments, or naturally occurring phenomena which cannot be explained by the theory. The above noted intrigues, two branches of science—mathematics and philosophy—cannot explain some examples, like—the spinning of planets but not their moons—by the hypothesis of the particle with a point on which the classical theories' tenets rest.

The validity and strength of the introduced concept of particle with two or more geometrical points and their connections emerge into successful and simultaneous applications to the different branches of science, mathematics, and philosophy, clearly explaining the observed facts, and do not rely on conjectured belief. The concept of connection, when applied to the above noted subjects simultaneously, solves the intriguing issues and brings the various subjects under one roof.

First, in making the final presentation of an idea, the smart ancient Greek intellectuals looked not just for the present, but also for the future. Their idea—"Philosophically, how small can materials split?"—is one of the most powerful, philosophical and purely speculative notions, having consequences not just at the time of Democritus, Epicurus, Heraclitus, Thales, and others as a basis for materialistic philosophy related to the smallest material element—the particle of an atom—which cannot be further divided, but had influenced Boyle—the founder of modern chemistry, Newton—the originator of mathematical physics, and in our time, Einstein—the father of theory of relativity. The philosophical idea of a particle, in turn, is immensely powerful in affecting the views of able researchers, including our contemporary working in various branches of mathematics and sciences. It is for this reason that a literary work deviating from the concept of particle with a point may seem to have little relevance to modern physics. But we needed to and have departed from it.

An argument of vision, a reason to have full potential, many before me believed in particle, and in this book, I started with a material particle, but added an extra association to the point of particle with other geometrical points to go beyond the theory of Newton's particle concept for planets and other rigid bodies; Maxwell's body concept for the electrons and charged particles; that turns out to be a slam-dunk shot to explain the spinning of the earth and electrons. These added geometrical points with the material particle point did maintain the Greek philosophers', Boyle's, Newton's, Maxwell's, Einstein's and others philosophy interest on how small the material can split. We do not need to answer the question: How small material can split to form an atom? Or the question: Is photon a particle or wave? Without addressing these questions, we presented and explained all the atomic elements of the periodic table.

Second, let us see how mathematics contributes to the number of atomic elements of the periodic table. By using mathematics and through connections, we will see that there are only 118 atomic elements in the periodic table, and no more than 118 are possible.

The seven rows of elements of the periodic table appear in two different types of series. First, the rows 1, 2, 3, 4, 5, 6, and 7 of the periodic table deal with natural, positive integer numbers associated with the electrons. The first row has only 2 elements with maximum of 2 electrons. This is due to the fact that the electrons appear as particle, 0 dimensions, but it has a property of positive and negative charges as Coulomb had originally observed. Thus, there are only 2 particle electrons possible for the 1st row.

Similarly, the dual property appears in the remaining 6 rows of elements in two different forms. First related to the positive and negative electrons, as is the case with 1st row; also the electrons are acting more than a particle and so have front and back orientations with respect to its material and the space in which they are moving. Thus, the remaining six rows to have a set of two rows with similar geometrical properties forming 3 pairs of two rows. For example: the 2nd and 3rd rows have electrons with 0 and 1 dimensions; 4th and 5th rows have electrons with 0, 1 and 2 dimensions; 6th and 7th rows have electrons with 0, 1, 2, and 3 dimensions. The 0 dimension electron has no pair, so the 1st row appears as a lonely unpaired row.

Thus, the seven rows turn into an arithmetic progression series (in which the difference between two successive terms is constant; for this case the constant is 2), for the four pairs—1, 3, 5 and 7, having electrons with the similar properties. The series permit us to calculate the number of elements in a specific pair of rows that is obtained by adding the number of elements in the previous series number, which can be represented as follows. The first row has only $2*(1) = 2$ elements, but with single row. So when we sum up the possible elements in the row we need to multiply the total number of elements for 1st row with 1. Each of the 2nd and 3rd rows has a total of $2*(1+3) = 8$ elements. For the 4th and 5th rows, each has a total of $2*(1+3+5) = 18$ elements. And finally, 6th and 7th rows, each has a total of $2*(1+3+5+7) = 32$ elements. The sum of this number for 4 pairs is equal to $(1*2+2*8+2*18+2*32 =)$ 118 elements in 7 rows. Note that the 1st row is a lonely (1) row with 2 elements and not 2 (rows) as is the case with the remaining 6 rows.

The same 7 rows can also be seen as four pairs: 1st row, 2nd, and 3rd rows, 4th and 5th rows, and 6th and 7th rows forming a series, respectively, of 0, 1, 2, 3 dimension electrons, with its material having a minimum of 1, 2, 3, and 4 points of geometrical connections in the space. In each of these 4 pair of rows, electrons are positive and negative creating 2 sets of elements.

To permit negative and positive electrons for each pairs, the number of elements in each row appears as square of the minimum number of geometrical points of connection. Thus, the 1st row has $2*(1)^2 = 2$ element; the 2nd and 3rd rows each has $2*(2)^2 = 8$ elements; the 4th and 5th rows each has $2*(3)^2 = 18$ elements; the 6th and 7th rows each has $2*(4)^2 = 32$ elements. Recall the 1st row is a single row while reaming rows are in pair of 2 rows. Thus, the sum of these numbers for 4 pairs is equal to $(1*2+2*8+2*18+2*32 =)$ 118 elements in 7 rows

In the 6th and 7th rows, electrons are with time-dependent connections. The time dependent orbits of electrons in the 7th row are larger, in comparison to the elements of the sixth row, and so the last $[2*(5+7) =]$ 24 elements are highly unstable. Due to this reason, in nature and in the earth's crust we observe only $(118-24 =)$ 94 elements in pure form or in combination of these elements. The remaining 24 elements are highly unstable, do not appear in nature, are synthesized in a laboratory, are short lived, and cannot possibly be detected in the earth's crust. The periodic table consists of 94 natural long-lived and 24 laboratory-made short-lived elements.

Also, the material of electrons has up to 4 points of geometrical connections in their orbits around the nucleus. As noted above, these material points of connections permit up to 118 elements. If we want to consider, or plan to create an additional number of elements in a laboratory, we cannot do so as the maximum permissible connections associated with 4 points connected electrons are already exhausted. One cannot add additional geometrical points of connection to the material of electrons. No more than 118 elements are possible.

Third, there is a consensus among the physicists about the need to study the laws of attraction (and repulsion) of one particle against the other is a serious matter to understand, but there is no clear, comparable consensus about how these laws constitute to explain the motions of these particles taking place in space and time.

Kepler's, Galileo's, and Newton's laws of motion have demonstrated details of the orbital motion of planets and other rigid bodies, when considered as particle. This book demonstrates their limitations. These laws, in particular, could not explain the observed spin of the planets. We explained why the earth spins and why the moon does not, when both of them are under gravitational attraction, as they both have different number of points of connections.

Similarly by adding points of connections associated with the material particles, with the novice idea, we have demonstrated that electrons in the atomic structure moved from a particle structure to the curve, surface, and solid structures in the atomic elements. This is different, and may be despised to many conventional physicists, than that form the established classical concept of particles in orbital motion.

Each row of the periodic table ends with all the possible orbits fully occupied with and saturated by two electrons in each possible and the end up forming noble elements. The noble elements are inert gases, and are located in the eighth columns of the periodic table. A noble gas of the previous row appears into the center of the next rows elements. For example, helium element of the first row is at the center of all elements of the second row, and so on.

Fourth, as noted in the item three above, the last elements—the noble gases—fall in the eighth column of the periodic table and have common properties. Similarly, there are other elements with common properties that are observable among elements of two or more rows of the same column. Why had these elements observed common chemical properties are not obvious when we follow the current books on chemistry or quantum mechanics? The common properties follow from the similarity of structure of electrons and connection with the nucleus.

In the first row of the period table, there are only 2 elements with one orbit. In each element, the electron acts as particle. When the single orbit is occupied by two electrons, it forms a helium element. Helium element forms a noble gas and is at the center, at the heart of all elements.

Similarly, the atomic elements with all saturated orbits by all possible shaped electrons form the noble gases. For seven rows, there are seven noble elements, except the last one is highly unstable and short lived. All elements are noble are located in eighth column. These elements are: helium (He), neon (Ne), argon (Ar), krypton (Kr), xenon (Xe), radon (Rn), and the recently discovered element—Ogannesson (Og), have well defined structures and common geometrical characteristics.

We will talk about the common properties appearing in the remaining six rows. In each row, the first two columns are with electrons having a point structure property, while the last six columns are with electrons having a curved structure. We discussed about the eighth column's elements, the noble gases, in the above paragraph.

In the first column of the periodic table, there are six elements with the center of the element is a noble gas from the previous row, and only has one orbit with a single particle electron. These elements are: lithium (Li), sodium (Na), potassium (K), rubidium (Rb), cesium (Cs), and francium (Fr). They have alkali property, as they are known as Alkali metal elements. Alkali metal elements are relatively reactive to water forming alkali solution, like sodium-hydroxide.

The second column has also six elements with a single orbit, but two particle electrons. The single orbit is saturated with 2, a possible number of particle electrons. These elements are: beryllium (Be), magnesium (Mg), calcium (Ca), strontium (Sr), barium (Ba), and radium (Ra). These have alkaline properties, and are known as alkaline earth metal elements. Alkaline earth metal elements are relatively stable.

The seventh column has six elements with one less electron in the 3 possible orbits of linear shaped electrons (just before having the last electron in the orbit to turn into a noble gas). The element 117 of the period table is made in laboratory and is unstable. As noted above, the last 24 elements are made in lab and are short-lived. We will not repeat this comment for the elements of the seventh row and focus on the five rows only. So, there are 5 elements in the 7th column. These 5 elements are: fluorine (F), chlorine (Cl), bromine (Br), iodine (I), and astatine (At), and are known as halogen elements.

The remaining elements of the 2nd through 6th rows and in the third through the sixth columns are with curved shaped electrons. The elements in each column have common properties depending upon how many electrons are there in possible orbits. There are 25 such elements. One can find the common properties of related to these elements in any advanced chemistry book.

The first 18 elements of the first three rows of the periodic table use the inherent property of four points of connections electrons and originate eight columns. These elements carry their common chemical properties into the remaining 4 rows of elements, but its reverse is not true. Thus there are only 8 columns of the periodic table, and we will name them: "Fundamental Columns."

For the other columns, we will name them as "Derived Columns." The elements in the derived columns do not have elements that have common properties of the fundamental columns.

For the fourth and fifth rows of the periodic table, the common properties elements with the particle and curved structured electrons appeared into the fundamental columns; there are additional 10 elements with surface structured electrons. Corresponding to these 10 elements, there are 10 sets of elements that are in the corresponding locations of the derived columns. Of these 10 sets, three elements have common properties in the derived columns. For example, copper (Cu), silver (Ag), and gold (Au), or zinc (Zn), cadmium (Cd), and mercury (Hg), and so on are sets of three elements have common properties. The elements with surface electrons of the seventh row are not included in this list as they are in laboratory-created elements and are short-lived.

The elements of the lanthanide and actinide series are with volume structured electrons and do not have common property elements either in the derived or in the fundamental columns.

## Last Comment:

How far we have come and what are its consequences are based on the concept of particle having additional geometrical and material connection points. I am not just talking about the (gravitational) particles of the Newtonian classical mechanics. I am talking about all including those of Newtonian particles, plus the electrons of Maxwell's electrodynamics, and the one appearing in the atomic elements of the periodic table. We easily resolved the four unrelated issues, and other similar properties presented in the previous chapters.

In the understanding of the motions of electrons, appearing in Maxwell's electrodynamics and in the elements of the periodic table, have been definitely and indirectly replaced by Faraday, a man with no formal education and in the beginning of his life he had to work as a low level employee in the chemistry lab of Davy in Cambridge, England, but was able to introduce a simple, but a powerful concept of magnetic tubes. This fable looking magnetic tube concept has three dimensional geometrical connections associated with electrons, is far stronger and better representation than the one we have tried and used to understand the electric and magnetic fields.

One can understand the strength of these connections, similar to the one presented by Faraday, in the form of applications to the similar applications we observe in our daily lives, without giving a second thought on the photons and with its controversial characteristics of particles or waves. A simple offering of a lighted candle for any occasion with associated circumstance—may be good or may be bad, like an offering to celebrate the good occasion or an offering for sympathy to the lost loved one, or the one having a candle light romantic dinner, or an offering to God for the theist, and so on—establishes a larger connection with nature, the universe. This is one of the easy ways to see the photons, light. Based on the connection presented here in this book for planets and electrons, it is possible to see a clear extension of these connections (not discussed in this book), when the electrons are excited, and when moved back (redux) into the original steady state orbital motion, it releases light particles—the photons. How interesting it would be to learn and to understand more about these photons!

# Apparatus and Method for Showing that A Magnetic Field Produces a Couple and Not a Force

## Why? +

Because motions of unpaired electrons produce magnetic field of equal and opposite forces in the neighborhood originates a couple. +

# Apparatus and Method for Showing that A Magnetic Field Produces a Couple and Not a Force

## United States Patent:    Joshi

Patent No:          US 007,471,083, B1

Date of Patent:     Dec. 30, 2008

Inventor:           Ramesh L. Joshi, Fremont, California, USA

Notice (*):         Subject to any disclaimer, the terms of this patent is extended or adjusted under35 U.S. C. 154(b) by 0 days, *Cited by examiner

Application No:  11/972,566

Filed:              Jan 10, 2008

Int. Cl.            G01R 33/12 (2006.01)
                    G01R 33/02 (2006.01)

U.S. Cl.            324/244; 324/260

Field of Classification Search
                    324/173-174, 324/228, 244, 260-261, 250/306
                    See application file for complete search history

| Reference Cited: | | | | |
|---|---|---|---|---|
| 1,735 | A | 8/1840 | Cook | |
| 1,143,529 | A | 6/1915 | Garretson | |
| 2,650,344 | A* | 8/1953 | Lloyd | 324/232 |
| 4.293,815 | A | 10/1981 | West et al. | |
| 4,414,285 | A | 11/1983 | Lowry et al. | |
| 5,681,987 | A* | 10/1997 | Gamble | 73/105 |
| 6,794,863 | B2 | 9/2004 | Hatanaka | |
| 6,977,505 | B1 | 12/2005 | Rosenquist | |
| 7,038,450 | B2 | 5/2006 | Romalis et al. | |
| 2006/0219324 | A1* | 10/2006 | Ozawa | 148/100 |

* Cited by examiner

Primary Examiner   Bot LeDynh
Attorney, Agent or Firm – Craig M. Stainbrook; Stainbrook & Stainbrook, LLP

*Ramesh L. Joshi, Ph.D., P.E.*

# Abstract of the Disclosure

An apparatus for observing magnetic phenomena including a nonferrous planar platform, a cylinder disposed upright on the platform, magnets placed on the platform at the lower end of the cylinder, specimen suspending means disposed at the upper end of the cylinder, ferrous and nonferrous specimens selectively connected to the specimen suspending means when the other specimen is not so connected, a shaft disposed through opposing holes in the side of the cylinder, and biasing means disposed inside the cylinder between the specimen suspending means and the shaft.

**14 Claims, 4 drawing sheets**

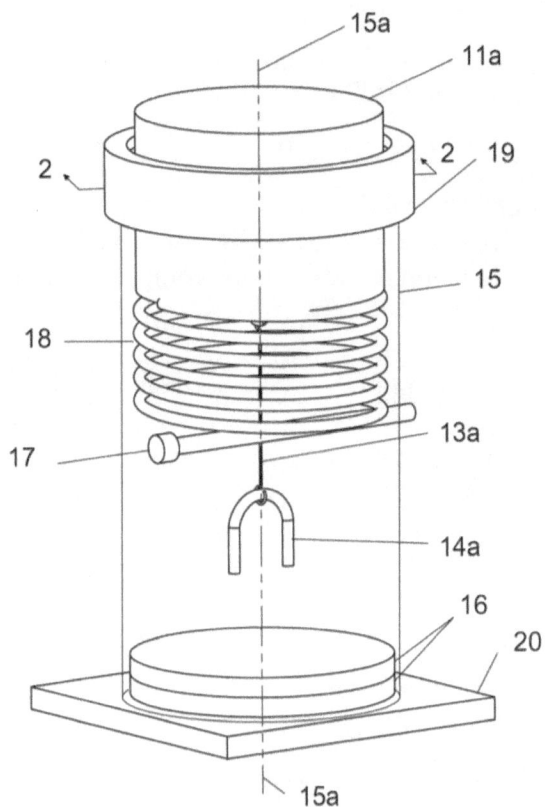

U.S.. Patent      Dec.30, 2008      Sheet 1 of 4      US 7,471,083 B1

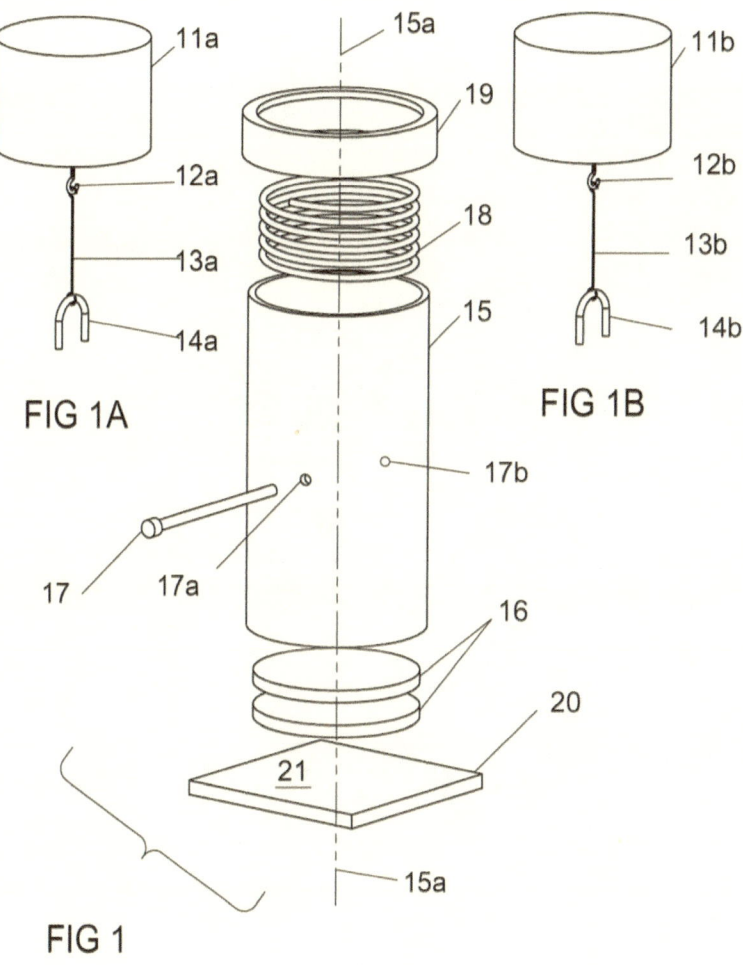

FIG 1A

FIG 1B

FIG 1

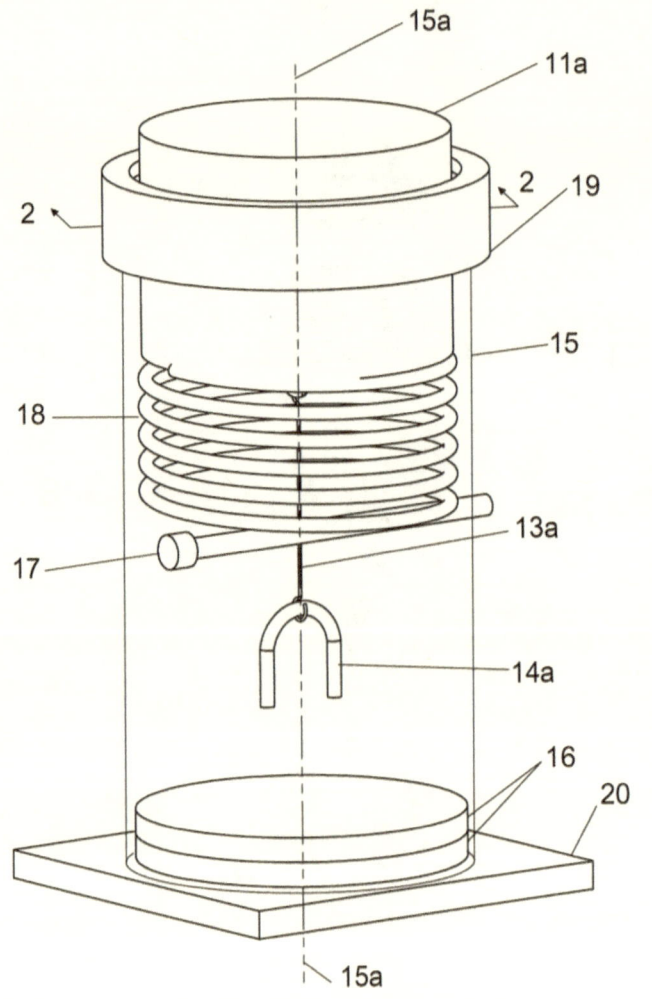

FIG. 2

176

U.S.. Patent          Dec.30, 2008          Sheet 3 of 4          US 7,471,083 B1

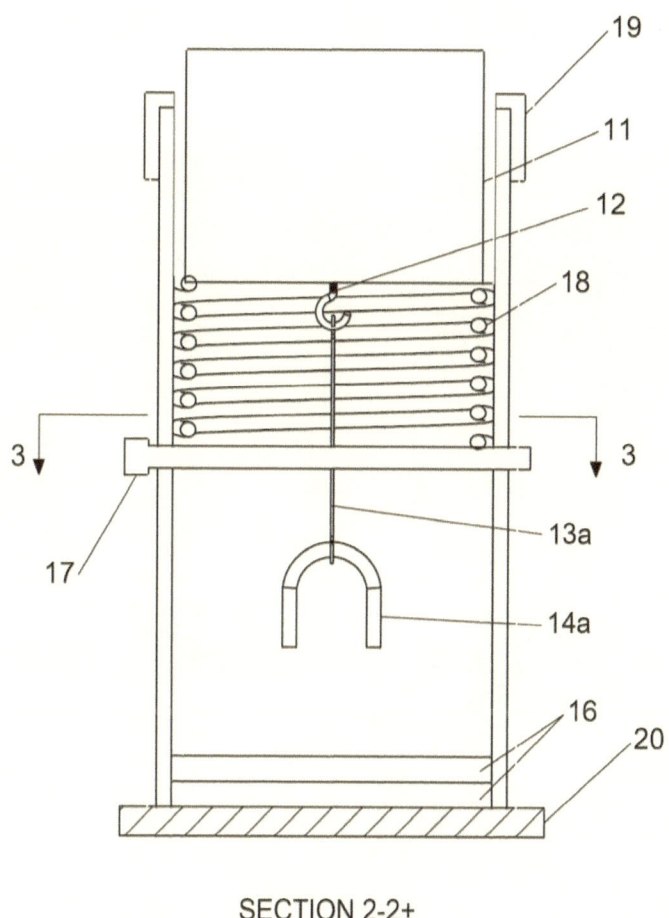

SECTION 2-2+

FIG. 3

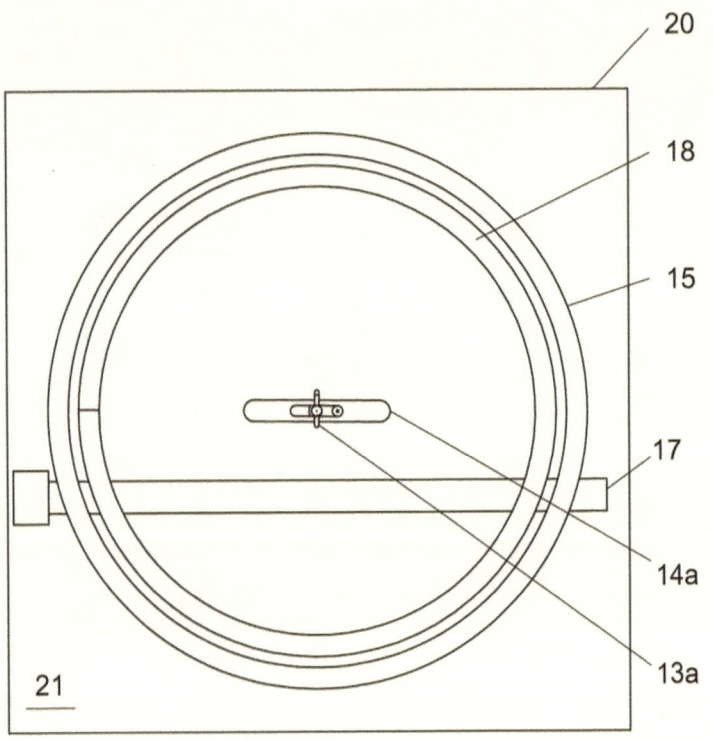

SECTION 3-3+

FIG. 4

# Apparatus and Method for Showing that A Magnetic Field Produces a Couple and Not a Force

CROSS REFERENCES TO RELATED APPLICATIONS
   Not applicable. The present application is an original and first-filed United States Utility Patent Application

STATEMENT REGARDING FEDERALLY SPONSORED RESEARCH OR DEVELOPMENT
   Not Applicable

THE NAMES OR PARTIES TO A JOINT RESEARCH AGREEMENT
   Not Applicable.

INCORPORATION-BY- REFERENCE OF MATERIAL SUBMITTED ON COMPACT DISC
   Not Applicable.

SEQUENCE OF LISTING
   Not Applicable

## Background of the invention

1. Field of Invention

The present invention relates generally to magnetometers, electromagnetism, and the study of magnetic fields and more particularly to an apparatus and method for the observation of magnetic phenomena, including the observation that the magnetic field produces a couple.

2. Discussion of Related Art

The apparatus described in the instant disclosure demonstrates that the magnetic field produces a couple and not a force, as conventionally believed and currently taught in physics. In this vein, a "couple" is understood to mean a pair of forces acting on parallel lines, equal in magnitude, opposite in directions, and at a finite distance that is non-zero and is known in physics as the "arm" of the couple. The apparatus of the present invention deals with the qualitative and quantitative measurements of magnetic field. The measurement criteria are a function of both displacement and time consumed per displacement. The invention thus enables measurement of the most desired perpendicular magnetic field as opposed to existing methods which work on the force due to magnetic field; existing methods are limited by measuring only the horizontal field, as the horizontal field produces a force which is easily measurable. Furthermore, the existing method deals with the magnetic forces. By contrast, the apparatus of the present invention measures the couple developed by the magnetic field. This method of measurement has been developed in view of quantum mechanics, but is not disclosed in either classical physics or known treatments of quantum mechanics.

Magnetic fields have been known for millennia. Current theories in physics view the magnetic and electric fields as different aspects of a single phenomenon called electromagnetism. Reducing electric and magnetic fields into a single electromagnetic field does not reveal, but rather conceals, the fundamental properties and differences of the fields of these phenomena. The geometrical characteristics of these three fields when experimentally observed have completely irreconcilable orientations with respect to their surroundings, and have different geometrical relationships to space, time and matter.

In 1820 Hans Christian Oersted discovered that an electric current produces a magnetic field causing a deflection in a magnetic needle when it flows over the needle. In 1831 at the Royal Institute of London, Michael Faraday experimentally observed that a magnetic field induces an electric current in a (copper) coil. At roughly the same time and independently, Joseph Henry in America and Heinrich Lenz in Russia discovered the same experimental results related to electric and magnetic fields.

In 1832 Carl Friedrich Gauss built upon these discoveries to engineer a magnetometer consisting of a bar magnet suspended from a gold thread. Using an improved apparatus of the same essential design, he was eventually able to measure the Earth's magnetic field.

Gauss also introduced a law related to magnetic flux. The Gauss magnetic flux law states: "The net magnetic flux through any (real or imaginary) closed surface is zero." The magnetic flux through an element of area perpendicular to the direction of magnetic flux is a measure of magnetic quantity equal to the product of the magnetic field and the area element scalar.

In 1865, James Clark Maxwell unified the experimental results of Coulomb, Gauss Oersted, Faraday, Ampere, and others into a set of four equations, known as the Maxwell's Electromagnetic Field Equations.

In one essential aspect, the present invention functions as a magnetometer similar to the one developed by Gauss, though it reveals residual magnetic properties neither observed nor reported either in early works or in current physical theories. Gauss's earliest magnetometer contained an element consisting of a bar magnet suspended from a gold thread. The present invention improves upon Gauss's magnetometer by introducing novel structures and features. These additional structures and features include, but are not limited to, a clear cylinder, the location of the magnets within the cylinder, and a suspended ferrous metal horseshoe. Additionally, a preferred embodiment of the present invention alters the shape of Gauss's bar magnet to that of a more effective horseshoe shape. While the prior art does utilize horseshoe magnets in other types of magnetometers, it does not suggest the use of a suspended horseshoe magnet, or ferrous metal horseshoe specimen as an improvement for Gauss's bar magnet. [See US Pat. No.1735, to Cook; and US Pat. No. 1,143,529 to Garretson.]

Furthermore, the present invention manually maneuvers a suspended ferrous material in a cylinder also containing the source of a magnetic field. Gauss's magnetometer, on the other hand, was housed in a large room and was designed to detect a magnetic source outside the housing structure, most famously, the Earth's magnetosphere. Furthermore, unlike Gauss's early magnetometer, which was not easily assembled, moved, or resembled, the apparatus of the present invention is transportable and easily disassembled and reassembled.

Many other types of magnetometers are found in the prior art. Well known examples of later developed magnetometers include the fluxgate, Overhauser, and atomic magnetometers. [Cf., US Pat. Nos. 4,293,815; 6,977,505; and 7,038,450.] However, all of these later magnetometers utilize structures and innovations other than a manually maneuvered suspended ferrous material for detecting magnetic fields.

The attractive and repulsive behaviors of magnetic poles are presently treated as being similar to the phenomena related to electric charges. This similarity between the magnetic and electric fields ends only at the attractiveness and repulsiveness; it does not appear to be inherent in these fields, and it cannot be extended to an isolated magnetic pole, which does not exist in the way that an isolated electric charge exists. If a bar of magnet is broken into two pieces, two isolated north and south poles do not occur; rather, the pieces maintain distinct north and south poles. If the process of breaking is continued, isolated north and south poles are still never created.

One of the fundamental properties satisfying the Maxwell's equations and the Gauss flux law is that the magnetic field is space dependent, which is not the case for an electric field. A magnetic field is produced due to the motion of an unpaired electrons in an atom (of ferrous) matter with specific orientation. To produce a magnetic field, there must be a minimum of two unpaired electrons in an atom (or molecule), both at a finite distance in the atomic/molecular orbit, and those unpaired electrons must have specific orientations which nullify the electric fields of the two electrons. In the magnetic field, the space separation of the minimal two unpaired electrons and their space separation dependency cannot be reduced to zero. This is the unique property of the magnet and magnetic phenomena. There is no similar space dependency in the electric (and gravitational) field(s).

To show a geometrical picture of the anomalous property of the magnetic field, we limit our discussion to two unpaired electrons, such as might appear in a nickel atom (which can be extended to more than two unpaired electrons without a loss of generality). These two unpaired electrons, at a finite distance and in motion, nullify the electric fields under certain conditions and produce local magnetic fields at the atomic level, with north and south poles orientations.

Under a normal condition, most unpaired electrons in an atom are unoriented but have a domain of magnetic boundaries with local magnetic fields randomly oriented and so, nullify the fields' effects, resulting in zero net magnetization. When an external magnetic field is applied to the material, boundaries between the magnetic domains move and produce an observable permanent magnetic field. The north and south poles are located along the applied field, creating a permanent magnet. According to the Gauss law, with no motion in the produced permanent magnet, the magnetic field lines do not start and stop at any point in the space, but form closed loops issuing from the North Pole and returning through the South Pole.

Now, the forces associated with the magnetic field lines in the neighborhood of one of the poles, say North Pole, are perpendicular to the field lines, are at the same distance apart as the electrons in the atomic structure. They also oppose each other and form a couple, like a spur gear structure. This structure is permanent for a permanent magnet.

As we move away from the poles, the couple structure gets weaker; the couple reduces to two forces, where the forces are opposites and are far away, appearing as pair of individual forces. These forces are weakest in the middle of the two poles of the magnet. However, when a permanent magnet is broken into two pieces, there appear new magnetic phenomena with newly added north and south poles and a force field structure associated with the poles.

A similar force structure also exists in material with two unpaired electrons. To easily observe the gear-and-tooth structure, a horseshoe specimen may be provided. When the horseshoe specimen is brought into proximity in front of a permanent magnet, it produces an induction in the specimen, and when given an up and down motion in the neighborhood of a pole, it rotates as if it is moving with a spur gear. The inventive apparatus reveals the couple force structure in the rotational motion of the horseshoe specimen.

# Brief Summary of the Invention

The present invention is an apparatus and a method for the study of magnetic phenomena. The novel apparatus of the present invention is an improvement Gauss's earliest magnetometer. It is easily portable and capable of detecting the magnetic fields of objects placed within a cylinder. Additionally, it does not require a connection to an external electrical power source.

In a preferred embodiment, the present invention includes a transparent cylindrical chamber with one or several magnets are placed at the lower end of the cylinder. A ferrous iron (or cobalt, or nickel, or any material with unpaired electrons) horseshoe is suspended from a dowel and lowered into the cylinder. A spring within the cylinder allows for controlled and smooth movement of the suspended horseshoe specimen.

As the iron horseshoe is maneuvered up and down the interior of the cylinder, the magnetic field of the magnets and the magnetic flux of the moving ferrous material create a magnetic induction causing the horseshoe to rotate. The strength of the magnetic field can be observed by the proportional rate of rotation of the horseshoe as the horseshoe is itself maneuvered at a constant rate. In this manner, the apparatus of the present invention is a novel magnetometer and demonstrates that the magnetic field produces a couple.

Other novel features which are characteristic of the invention, as to organization and method of operation, together with further objects and advantages thereof will be better understood from the following description considered in connection with the accompanying drawings, in which preferred embodiments of the invention are illustrated by way of example. It is to be expressly understood, however, that the drawings are for illustration and description only and are not intended as a definition of the limits of the invention. The various features of novelty that characterize the invention are pointed out with particularity in the claims annexed to and forming part of this disclosure. The invention does not reside in any one of these features taken alone, but rather in the particular combination of all of its structures for the functions specified.

There has thus been broadly outlined the more important features of the invention in order that the detailed description thereof that follows may be better understood, and in order that the present contribution to the art may be better appreciated. There are, of course, additional features of the invention that will be described hereinafter and which will form additional subject matter of the claims appended hereto. Those skilled in the art will appreciate that the conception upon which this disclosure is based readily may be utilized as a basis for the designing of other structures, methods and systems for carrying out the several purposes of the present invention. It is important, therefore, that the claims be regarded as including such equivalent constructions insofar as they do not depart from the spirit and scope of the present invention.

## Brief Description of the Several Views of the Drawings

The invention will be better understood and objects other than those set forth above will become apparent when consideration is given to the following detailed description thereof. Such description makes reference to the annexed drawings wherein:

FIG. 1 is an exploded perspective view of a preferred embodiment of the apparatus of the present invention;

FIG. 2 is a perspective view showing the inventive apparatus fully assembled;

FIG. 3 is a cross-sectional side view in elevation of the apparatus of FIG. 2; and

FIG. 4 is a top plan view of the apparatus of FIGS. 2 and 3.

## Detailed Description of the Invention

Referring to FIGS. 1 through 4, wherein like reference numerals refer to like components in the various views, there is illustrated therein a new and improved apparatus to demonstrate that a magnetic field produces a couple and not a force. The apparatus is generally denominated **10** herein. The simple assembly includes a hollow cylinder **15** placed in an upright position on a nonferrous base or platform **20**, said base having a substantially planar upper surface **21**. The cylinder has a longitudinal axis **15a** and supports a specimen suspending means **11a/11b**, **12a/12b**, at least one shaft **17**, a ferrous specimen **13a** or, alternatively, a nonferrous specimen **13b**, and a wire or string **14a/14b**. One or more sets of apertures **17a**, **17b**, are located on opposing sides of the cylinder for insertion of the one or more shafts. The shaft(s) **17** is/are of a length greater than the outer diameter of the cylinder. The apertures are offset from the center of the cylinder to allow the centered suspension of the wire or string. At least one, and preferably a plurality of magnets **16** is placed within the perimeter of contact between the cylinder and the platform. Furthermore, each magnet is placed in such a manner that its magnetic field is aligned perpendicularly to the upper surface of the planar platform.

The suspending means **12** in the preferred embodiment preferably comprises a fixed anchor, such as a hook attached to the underside of dowel or plug **11a/11b** generally at the geometric center of the dowel. In each embodiment the dowel is a nonferrous material such as wood, plastic, glass, and the like.

One specimen must consist of a ferrous material such as iron. The other specimen must consist of a nonferrous material such as copper. Each of the specimens preferably has a horseshoe shape. A line of some kind, preferably lightweight string or wire **13a/13b**, attaches the specimen to the suspending means **12a/12b**.

The shaft **17** is inserted through opposing apertures **17a/17b** in the side of the cylinder. A spring **18** or other biasing means is then placed into the interior of the cylinder and disposed in an upright position on the shaft so as to not hinder the insertion of the specimen into the interior portion of the cylinder. The spring supports the dowel to which the anchor or hook **12a/12b** is attached.

An annular cap **19** surrounds and further centers the dowel **11a/11b**. The cap is concentric with the cylinder and includes an aperture sufficient for the sliding insertion of the dowel to which the specimen suspending means is attached. When placed inside the cylinder, whether or not the cap is included, the geometric center of the dowel is substantially aligned with the longitudinal axis of the cylinder. In operation, therefore, the specimen supporting means suspends a specimen within the interior of the cylinder substantially along the longitudinal axis of the cylinder. The height of suspension is varied by manually manipulating the specimen supporting means (i.e., depressing the dowel and allowing it to spring back up).

A ferrous specimen **13a** is then maneuvered (via depression of the dowel) to a position as close to the bottom of the cylinder as possible without touching the magnets. The specimen is allowed to come to a complete rest. The specimen is then maneuvered via manipulation (release of depression) of the dowel in a vertical direction upwards at a rate sufficient to produce magnetic induction of a magnitude capable of rotating the ferrous specimen. The angular velocity of the rotation is proportional to the rate the specimen is maneuvered.

The ferrous specimen is then removed and the process is repeated with the nonferrous specimen **13b**. It too is allowed to come to rest, and the vertical position of the nonferrous specimen is then maneuvered at a rate similar to the rate used for the ferrous specimen. However, it will be noted that the identical movement produces neither a magnetic induction nor a rotation in the nonferrous specimen.

The hollow cylinder of the preferred embodiment includes a rigid clear plastic tubular cylinder with an outer diameter of 0.75 inches, an interior diameter of 0.50 inches, and a length of 5.75 inches. Two round planar magnets with a thickness of 0.187 inches and a diameter of 0.4966 inches are glued in place to the platform with their magnetic fields in unison and oriented perpendicularly to the platform. The cylinder is placed in an upright position over the magnets such that the magnetic fields are oriented with the center axis of the cylinder.

The ferrous specimen is preferably constructed of a 0.0125 inches diameter and 1.5 inches length iron wire in a horseshoe shape and attached to a specimen suspending means with string of nonferrous material and a diameter of 0.002 inches. The nonferrous specimen is constructed of a 0.0125 inches diameter and 1.5 inches length copper wire in a horseshoe shape and attached to a specimen suspending means with string of nonferrous material and a diameter of 0.002 inches.

The specimen suspending means consists of a 0.48 inches diameter 1.35 inches long wooden dowel with a stainless steel eye inserted into its surface at its axial center. A hook is constructed 0.55 inches diameter wire and has a 0.226 inches diameter eye. The string portion of the specimen is then attached to the hook of a specimen suspending means.

A stainless steel shaft of 0.125 inches is inserted into the apertures at the side of the cylinder. A stainless steel spring of 0.4865 inches outer diameter, 0.3705 inches inner diameter with seven turns each of 1.375 inches in length is then installed in the cylinder. The spring both supports the specimen supporting means and permits the smooth motion of the specimen. The cap is then attached to the top portion of the cylinder securing the spring within the cylinder.

Above noted dimensions, or cylinder shape, or specimen shape can be changed without losses of observations as long as the intent and constrains are maintained, e.g. one can replace the circular cylinder with a clear square box.

As will be immediately appreciated from the foregoing description, in its most essential aspect, the apparatus of the present invention is an improved magnetometer for demonstrating that a magnetic field produces a couple. The essential elements of the preferred embodiments comprise (1) a nonferrous platform having a substantially planar upper surface for mounting a cylindrical chamber; (2) a cylindrical chamber disposed in an upright position on said upper surface of said nonferrous platform, said cylindrical chamber being rigid, hollow, and having a lower end and an upper end; (3) at least one magnet disposed at said lower end of said cylindrical chamber such that the magnetic field of said one or more magnets is perpendicular to said nonferrous platform; (4) a ferrous specimen fabricated from a ferrous material and a nonferrous specimen fabricated from a nonferrous material; (5) specimen suspending means disposed at said upper end of said cylindrical chamber for suspending either one of said ferrous specimen or said nonferrous specimen in a substantially centered position along the longitudinal axis of said cylindrical chamber; and (6) reciprocating means for moving said specimen upward and downward along the longitudinal axis of said cylindrical chamber.

Accordingly, while this invention has been described in connection with the preferred embodiments thereof, it is obvious that modifications and changes therein may be made by those skilled in the art to which it pertains without departing from the spirit and scope of the invention. Accordingly, the scope of the invention is limited only to the appended claims.

Further, the above disclosure is sufficient to enable one of ordinary skill in the art to practice the invention, and provides the best mode of practicing the invention presently contemplated by the inventor. While there is provided herein a full and complete disclosure of the preferred embodiments of this invention, it is not desired to limit the invention to the exact construction, dimensional relationships, and operation shown and described. Various modifications, alternative constructions, changes and equivalents will readily occur to those skilled in the art and may be employed, as suitable, without departing from the true spirit and scope of the invention. Such changes might involve alternative materials, components, structural arrangements, sizes, shapes, forms, functions, operational features or the like.

Therefore, the above description and illustrations should not be construed as limiting the scope of the invention, which is defined by the appended claims.

## Claims

What is claimed as invention is:
1. An apparatus for observing magnetic phenomena, including a force couple of the magnetic field, comprising:
    (a) A base having a substantially planar upper surface;
    (b) A cylinder disposed in an upright position on the upper surface of said base, said cylinder being rigid, hollow, having a lower end and an upper end, being open at least at said upper end, and having a pair of opposing apertures disposed in the side of said cylinder;
    (c) At least one magnet placed on said base at said lower end of said cylinder in a manner such that the magnetic field of said at least one magnet is perpendicular to said base;

(d) Specimen suspending means slidingly disposed at said upper end of said cylinder;

(e) A ferrous specimen fabricated from a ferrous material and a nonferrous specimen fabricated from a nonferrous material, either of said nonferrous specimen and said ferrous specimen selectively connected to said specimen suspending means when the other of said specimens is not connected to said specimen suspending means, whereby the specimen is suspending along the longitudinal axis of said cylinder;

(f) At least one shaft disposed through the opposing apertures in said cylinder; and

(g) Biasing means disposed inside said cylinder between said specimen suspending means and said shaft.

2. The apparatus of claim 1, wherein said cylinder is fabricated from a transparent plastic material.

3. The apparatus of claim 1, wherein said base is fabricated from nonferrous material.

4. The apparatus of claim 1, wherein each of the magnets of said at least one magnet is round and planar.

5. The apparatus of claim 1, wherein said ferrous specimen has a horseshoe shape.

6. The apparatus of claim 5, wherein said ferrous specimen is made of iron.

7. The apparatus of claim 1, wherein said nonferrous specimen is horseshoe shaped.

8. The apparatus of claim 7, wherein said nonferrous specimen is made of copper.

9. An apparatus of claim 1, wherein said biasing means is a spring.

10. A method for demonstrating that a magnetic field forms a force couple of the magnetic field, comprising the steps of:

(a) Providing a nonferrous planar platform, a rigid hollow cylinder placed on the platform in an upright position, the cylinder being open at both ends and having a pair of opposing apertures in the side of the cylinder, at least one magnet placed on the planar platform at the lower end of the cylinder in a manner such that the magnetic field of each magnets is perpendicular to the planar platform, a specimen suspension means disposed at the upper end of the cylinder, a ferrous specimen fabricated from a ferrous material and connected to the specimen suspending means, a nonferrous specimen fabricated from a nonferrous material and connected to the specimen suspending means when the ferrous specimen is not so connected, and means for reciprocating the specimens upwardly and downwardly along the longitudinal axis of the cylinder;

(b) Inserting the ferrous specimen into the cylinder and suspending it with the specimen suspension means;

(c) Manipulating the specimen suspension means to maneuver the ferrous specimen within the cylinder in an up and down reciprocating manner along the longitudinal axis of the cylinder and observing the movement of the ferrous specimen;

(d) Removing the ferrous specimen from the cylinder and disconnecting it from the specimen suspension means;

(d) Inserting the nonferrous specimen into the cylinder and suspending it with the specimen suspension means; and

(e) Manipulating the specimen suspension means to maneuver the nonferrous specimen within the cylinder in an up and down reciprocating manner along the longitudinal axis of the cylinder and observing the movement of the nonferrous specimen.

11. The method of claim 10, further including the step of using a horseshoe-shaped ferrous specimen to observe the perpendicular magnetic material as a couple.

12. An improved magnetometer for demonstrating that a magnetic field produces a couple, said apparatus comprising:

A nonferrous platform having a substantially planar upper surface for mounting a cylindrical chamber;

A cylindrical chamber disposed in an upright position on said upper surface of said nonferrous platform, said cylindrical chamber being rigid, hollow, and having a lower end and an upper end;

At least one magnet disposed at said lower end of said cylindrical chamber such that the magnetic field of said one or more magnets is perpendicular to said nonferrous platform;

A ferrous specimen fabricated from a ferrous material and a nonferrous specimen fabricated from a nonferrous material;

Specimen suspending means disposed at said upper end of said cylindrical chamber for suspending either one of said ferrous specimen or said nonferrous specimen in a substantially centered position along the longitudinal axis of said cylindrical chamber; and

Reciprocating means for moving said specimen upward and downward along the longitudinal axis of said cylindrical chamber.

13. The magnetometer of claim 12, wherein said reciprocating means comprises a dowel slidingly disposed at said upper end of said cylindrical chamber.

14. The magnetometer of claim 13, wherein said specimen suspending means comprises a fastener affixed to said dowel and a line connected to said fastener and to said specimen.

## To Observe Spin of Electrons +

In Chapters IV and V we noted that an electron has observable front and non-observable back-side (obverse) manifolds, except when it is in a particle state. As a particle, an electron has no dimensions, no front, no back, and giving rise only two atomic elements in the first row of the periodic table; we cannot, in theory, observe any spin out of the atomic elements in which electrons act as particles, which include the first row and all the elements of the first two column of the remaining six rows.

In practice, the spin of particle electron in hydrogen element is observed, known after the inventor Pieter Zeeman (1865–1943), as the Zeeman Effect of "duplexity phenomena" as two spectral lines are observed with a help from fairly very good spectroscopic equipment, and light spectra is passed through very strong magnetic field. This observation is done through photons and not through electrons. In this presentation, we will continue to focus to report on the spin of the electrons based on their physical properties and not on the spectral lines observed through spectroscopic equipment.

In the remaining atomic elements of the periodic table the electrons have curves, surfaces, and three dimensional manifolds that have observable fronts and non-observable back (obverse) sides, and are spinning, but are not easily observable simultaneously.

An absence of observational evidence of electron spin in atomic elements with the noted manifolds is neither an evidence of its absence, nor is the evidence of its presence; it demands further investigation and understanding of its geometrical characteristics and influence to and from the neighboring space in which they are moving.

The spin of electrons are theoretically perceived based on its four point connections, which led us to explain the spin of the earth, beyond the Newtonian theory presented in Chapter I. The spin of electrons cannot be (essentially and easily) observed, similar to the spin of the earth, or other Newtonian particles, as they are not even visible to the naked eyes; how can we easily observe its spin?

From the four point connection, the spin is an inherent—intrinsic —property permanently associated with the electron. The spin effect appears in motion of electrons, but it does not depend on the circumstance of the motion—such as velocity, kinetic energy, or angular motion that appears with Newtonian particles in motion.

The perceived electron spin was introduced in quantum mechanics to explain the observed dual spectral lines of light passing through a magnetic field interpreted having two energy states associated with motion of electrons. In this mindset to explain the observed facts related to light spectra lacks to represent the true intrinsic property (of geometrical connection) associated with the electron spin. This is tragic—as it appears and rises from a disheartening cloud of spectra of light and not associated with the intrinsic property of the electron spin—and it is understandable as it does not address the need to look into how the motion of electron takes place in its orbital motion in any atomic element or in free space. These observed facts of light spectra passing through magnetic field separating into two spectral lines, however, suggest that one should plan to consider having motion of atomic elements through a magnetic field.

In addition to the presentations in Chapters IV and V, and from quantum mechanics we know that the electrons in the atomic elements theoretically spin, and so the fact is tempting and it is physically possible to observe the spin though its existence evidence is not easily visible. For the excitement to observe the spin, we need to recall how nature has kept secrets in revealing the physical facts. We are forced to understand her activities and work in pursuit to reveal the secrets.

The earth and the moon are revolving under the influence of gravitational field, and yet we only observe the spin of the earth and not that of the moon. On the other hand, in comparison to a celestial body, the electrons are small and are hard to observe with the naked eyes when it is either at rest or in motion, and so it is hard, at a first attempt, to observe its spin unless we take into account of having influence from its neighboring space.

We should not presume that it will be easily possible to observe the effect of an electron's spin for all the atomic elements. Only the few elements will show the spin effect which has the necessary geometrical connections with the space in which they appear.

The electrons in atomic elements have particle, curve, surface, and three dimensional manifolds. It is hard to observe spin of electrons in particle manifolds. The curve and three dimensional manifolds have observable fronts and non-observable back-side (obverse); these manifolds spin while orbiting around the nucleus; it is hard to observe the spin of these manifolds or even recognize them in their time dependent orbital motion as the associated tangent and normal, respectively, remain on the same plane (surface) or on the same three dimensional spaces. And the space, time, and matter are absolute, and no spin is observable for these manifold-electrons.

The electrons with surface manifolds have normals and tangents that change with time in orbital motions. The direction of tangent to the (surface) manifold changes its direction with the change of geometry of the surface, while the normal to the surface remains perpendicular to the surface and so the change of normal is not noticeable in its orbital motions. Only the change in tangent and its associated (magnetic) field directions will permit the observation of the spin of surface manifold electrons provided the magnetic field is not dependent on time as the space and time in this case are not united as the electrons are moving slowly in the orbit.

There are 10 possible surface manifolds of electrons for the 4th to 7th rows of the elements in the periodic table. At the most 2 electrons can occupy an orbit permitting to have 5 possible orbits for the surface manifold electrons. The electrons first occupy one by one all of the five permissible orbits. The electrons in these 5 orbits have time dependent motion with a surface manifold. These 5 orbits' case is similar to the other manifolds of the electrons and it is not feasible to observe the spin.

## Unpaired electrons' spin in fourth Row:

In the fourth row of the periodic table, when the sixth electron occupies one of the 5 orbits of the element, the remaining four orbits have unpaired electrons in orbital motion with surface manifolds in the remaining 4 orbits. This is the case for an iron element in the fourth row of the periodic table. The spinning of these four electrons produce local magnetic field, independent of time, and do not demand to have the geometrical connection to be space-time. So, the spin of each of the four electrons in the iron element is possible to visualize and can be demonstrated under the local magnetic field.

When the seventh and eighth electrons occupy the second and third of the possible 5 orbits of the elements, the remaining third and second orbits have, respectively, 3 and 2 unpaired electrons in the orbital motions. These two cases are for cobalt and nickel elements in the fourth row of the periodic table. The spinning of these 3 and 2 electrons produce local magnetic field, independent of time, and do not demand to have geometrical connection to be space-time. So, the spin of each of these electrons in the atomic element of cobalt and nickel are also visible and can be demonstrated with local magnetic field.

When the ninth and tenth electrons occupy the fourth and fifth orbits of the elements, there are only one and all orbits, out of 5 orbits, respectively, with one and two electrons. In both these cases there is no orbit with the unpaired electrons. These cases are for copper and zinc elements and have no magnetic field possible, and cannot be considered to demonstrate the spin.

## Unpaired electrons' spin in fifth Row:

Similarly, in the 5th row, there are ruthenium, rhodium and palladium, respectively, which have 4, 3, and 2 unpaired electrons producing magnetic fields in the neighborhood of the atomic elements. The five orbits of electrons in the fifth row are larger than that of the fourth row. Thus, the unpaired electrons in fifth row produce weaker magnetic fields compared to the one produced in the fourth row. However, when used to make an alloy out of these elements, it provides stronger magnetic field and this can be observed well.

## Unpaired electrons' spin in 6th and 7th Rows:

In the 6th and 7th rows, respectively, there are osmium, iridium, platinum, and hassium, meitnerium, damstadtium, which are elements with 4, 3 and 2 unpaired electrons. These elements surface manifold electrons appear after the volume manifolds having much larger orbits compared to the 4th and 5th row elements. These electrons are unpaired, but do not originate (easily) observable magnetic field in the local neighborhood due to the size of their orbits. One cannot see the spin out of these elements either.

## To observe Electron Spin:

Electrons in the 118 elements of the periodic table spin in their orbits and locally produce electromagnetic fields. But there are only three atomic elements having electrons with surface manifolds that are possible to observe spin in their orbits. Why? Because these three elements unite with the local space and are independent of time to produce local magnetic field. This is not feasible for other electrons that have surface manifolds as their orbits are either large or do not have unpaired electrons and remain time dependent in their motion.

The above noted three elements may not have locally observable or physically measurable magnetic properties, but when moved in the magnetic field, the material spins on its axis to demonstrate spin of electrons, as it was the case of Zeeman Effect for the light spectrum.

For easy observation of an electron spin, one can consider the material having the maximum number of unpaired electrons (4) in the element. This material is iron. A (symmetrical) speck or a small plate of iron, referred as specimen, will have a magnetic field at the atomic level in all directions, reducing the resultant magnetic field to zero. To easily observe, to appreciate, to envisage, and to explain the electron spin we will discuss the case of the nickel element that has 2 unpaired electrons that permit to comprehend it in analogues to gear model or "sprockets and chain" model, with an adaptation to the characteristics to the magnetic field. We will only discuss the case of the gear model.

Let us recall that each atomic element of nickel has two unpaired surface manifold electrons orbiting on the boundary of its structural shell and spinning around its own axis each producing local electromagnetic field, but their resultant produce a local magnetic field, which is independent of time.

As noted in Chapter V, each electron with surface manifold has connection with a point outside of the electron material and, in its motion, time is absolute. Out of all these exterior electrons of the atomic elements of the nickel specimen we focus here on the two unpaired surface manifold electrons.

The unpaired electron of each element has connection with the space outside of the specimen of the material that originates a surface, interpreted as representing a tooth of a gear. As an electron spins in its orbit, its connection to the exterior point remains, as the electron has surface manifold mode maintaining its tooth structure. As the electron rotates in its orbit, the gear tooth deforms, but maintains its tooth surface structure until the second unpaired electron complements its deformation and maintains the gear structure, including the local magnetic structure of the nickel element. This is true for all elements on the same plane and entire surface of the of the nickel specimen. These exterior elements with unpaired electrons create the external spur (driven) gear with tooth being perpendicular to the axis of rotation along which we suspend the symmetrical specimen, as shown in the Figure 3 of the Appendix.

Let us move the symmetrical nickel specimen in open space, out of the magnetic field, up-down, and observe that it does not spin. Now, let us move to the same specimen up-down in the magnetic field and observe its spin as it moves down or it moves up; the spins are in opposite directions. This is similar to the case of two gears motion, driving the gear in one or the other directions. The needed forces to spin the specimen comes from the variable magnetic field of the permanent magnet with gear structure and the spinning of unpaired electrons originate the gear structure of the local magnetic field. In this spin there is no influence from the vertical motion of the specimen as observed outside of the magnetic field.

The exterior gear analogy of the nickel specimen can be extended, without loss of generality, to a permanent magnet with surrounding space filled with lines or tubes of a magnetic field, as threading or filling the area of the space, as Faraday had portrayed.

Consider the space is field with the tubes of a magnetic field. If we take a cross section of the magnetic field tube, it originates equal and opposite forces in the horizontal plane, perpendicular to the axis of spurs, to rotate a horseshoe and a nickel specimen. Thus one can picture that the space is filled with interior or exterior spur gears filling the space over the magnet originating the interior and exterior spur gears.

This representation of the magnetic field tube filling the space fundamentally differs from the concept of Newtonian mechanics as it portrays simultaneously of having interior and exterior gears in the same space, when time is still maintained to be absolute as Newton had considered, while the space unites with material and participates on equal footing with the matter in its motion. And the unpaired electrons reveal its spin in an up-down motion in the magnetic field.

These observations clarify the structure of Faraday's magnetic tubes of forces. These tubes are smooth surfaces when observed from outside, but have tooth and gear structures inside, originating equal and opposite forces, such that they meet together inside and fill the entire space. This structure is visible only when a specimen with unpaired electrons is moved up and down. This is an extension and clarification on the structure of the Faraday's tubes of forces.

+Added items to the awarded patent.

# Acknowledgements

As one who has studied at various levels of research work on the foundation of the subject, I am aware of the negative attitude of so many towards the subject of addressing to amend the foundation. There are many reasons for this, one of them, no doubt, is raising a red flag against the establishment who prefer not to change what is established; so the establishment will not support an individual to have a new concept to see that differences exist in physics. But there are many who support the one who is working to simplify, and to solve the mysteries of the dry subject; especially, I received that from my professors whose names appear in the Preface.

The needed moral support to write this book came from family-friends—Ramsey Maker, Howard and Pat Davis, Tom McGaw, Abdullah Kazi and Padmanabh D. Lele; illustrative drawings from—Kamlesh Panchal and Sina Dadras; encouragement from family members—Nikhil R. Joshi, Ashvin L. and Pat A. Joshi, Uren K. and Lopa U. Bhatt, and Darshana Keyur Bhatt. Last, but definitely not the least, my gratitude goes to my wife, Sudha R. Joshi, for her tolerance to my absentmindedness during the conceptual development, and invaluable support in preparing the manuscript of the book in bringing it to a reality.

# Index

# Notes